AQA KS3

Science 1

Extend

PRACTICE BOOK

Cliff Curtis
Deborah Lowe
Owen Mansfield

HODDER
EDUCATION
AN HACHETTE UK COMPANY

Hachette UK's policy is to use papers that are natural, renewable and recyclable products and made from wood grown in well-managed forests and other controlled sources. The logging and manufacturing processes are expected to conform to the environmental regulations of the country of origin.

Orders: please contact Hachette UK Distribution, Hely Hutchinson Centre, Milton Road, Didcot, Oxfordshire, OX11 7HH. Telephone: +44 (0)1235 827827. Email education@hachette.co.uk. Lines are open from 9 a.m. to 5 p.m., Monday to Friday. You can also order through our website: www.hoddereducation.co.uk

First published in 2017 by
Hodder Education,
An Hachette UK Company
Carmelite House
50 Victoria Embankment
London EC4Y 0DZ

www.hoddereducation.co.uk

Impression number 10 9 8 7 6 5 4 3

Year 2022

Cover photo © Laurel/Alamy Stock photo

Typeset in 11/14 pt Vectora 45 Light by Integra Software Services Pvt. Ltd., Pondicherry, India

Printed by CPI Group (UK) Ltd, Croydon, CR0 4YY

A catalogue record for this title is available from the British Library.

ISBN: 9781510402508

Contents

Find the answers at www.hoddereducation.co.uk/AQAKS3Science

1 Speed

» Calculating speed

Worked example

A sprinter runs 400m in 50 seconds.

a) What is her average speed?
b) How long would it take her to run 1 km at this speed?

a) Using the average speed equation:

$$\text{average speed} = \frac{\text{distance travelled}}{\text{time taken}} = \frac{400\,m}{50\,s} = 8\,m/s$$

b) Rearranging the equation and remembering that 1 km is 1000 m:

$$\text{time taken} = \frac{\text{distance travelled}}{\text{average speed}} = \frac{1000\,m}{8\,m/s} = 125\,s$$

> **Hint**
>
> If you want to find the distance travelled or the time taken, you have to rearrange the equation.

Know

1 What is the average speed of a dog that travels 60m in 5s?

2 What is the average speed of a bicycle that travels 50km in 2h?

3 What is the average speed of a rocket that travels 100km in 25s?

4 Copy and complete the table below. The first two rows have been done for you. Remember to include units.

Speed	Distance	Time
5 m/s	100 m	20 s
8 km/h	4 km	30 min
	48 m	4 s
	125 m	25 s
2 m/s	8 m	
24 m/s		10 s
40 km/h		3 h
	20 km	5 h
15 km/s	90 km	
3 m/s		1 h

Apply

1 A boy walks 200m in 40 seconds.

   a) What is his average speed?

   b) How far would he walk in five minutes at this speed?

   c) How many minutes would it take him to walk 2.5km at this speed?

2 The speed limit on busy roads in the UK is 30 miles per hour.

   a) How long would it take a car travelling at this speed on average to travel 90 miles?

   b) How long would it take a car travelling at this speed on average to travel 10 miles?

   c) How far would a car travelling at this speed on average travel in 45 minutes?

Extend

1 A girl starts from stationary and walks 300 m around the block. It takes her one minute and 45 seconds.

   a) What is her average speed?

   b) Explain why she must at some point have walked at a speed that is faster than this.

   c) How long would it take her to travel 1.8 km at this speed in minutes?

» Finding speed from a graph

Worked example

The distance–time graph below shows Izzie's journey to the corner shop.

1 Between what times is she
   a) travelling at a steady speed
   b) stationary
   c) travelling the fastest?
2 How far does Izzie travel in total?
3 What is Izzie's speed between 40 s and 70 s?
4 What is Izzie's average speed for the whole journey?

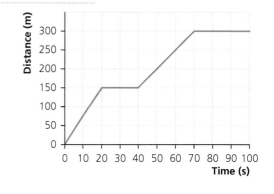

1 From the graph we can see that:
   a) Izzie is travelling at a steady speed between 0 and 20 seconds and between 40 and 70 seconds, because these parts of the graph are straight diagonal lines.
   b) Izzie is stationary between 20 and 40 seconds and between 70 and 100 seconds, because they are the horizontal parts of the graph.
   c) Izzie is travelling fastest between 0 and 20 seconds, because this is the steepest part of the graph.
2 300 m; this is the highest value reached by the graph on the y-axis, which shows distance.
3 Speed is calculated using the gradient.

   $$\text{gradient} = \frac{\text{change in } y \text{ value}}{\text{change in } x \text{ value}} = \frac{300 - 150}{70 - 40} = \frac{150}{30} = 5 \text{ m/s}$$

4 This is calculated using the total distance and the total time taken:

   $$\text{average speed} = \frac{\text{distance travelled}}{\text{time taken}} = \frac{300 \text{ m}}{100 \text{ s}} = 3 \text{ m/s}$$

Hint

Gradient is calculated by dividing the change in y-axis value by the change in x-axis value.

Know

1 Look at the four distance–time graphs below.

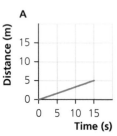
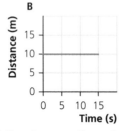
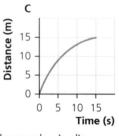
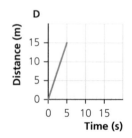

a) Choose one of the following captions for each graph a)–d):

| fast steady speed | slow steady speed | accelerating | stationary |

b) For the two steady-speed graphs, calculate the speeds that are represented.

Apply

1 Look at the graph of an elephant's journey.

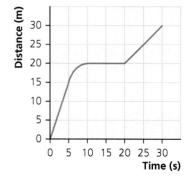

a) When is the elephant travelling at a steady speed?

b) When is the elephant stationary?

c) When is the elephant accelerating?

d) How far does the elephant travel in total?

e) What is the elephant's speed between 0 s and 5 s?

f) What is the elephant's speed between 20 s and 30 s?

g) What is the elephant's average speed for the whole journey?

2 Draw a distance–time graph for the following journey ride. Your *x*-axis should go up to 10 seconds and your *y*-axis should go up to 10 metres.

• The object travels 6 m in the first 3 s.

• It is then stationary for 2 s.

• It then travels 4 m in the next 5 s.

Use the graph to calculate the speed of the object during each part of its journey, and then the average speed for the whole journey.

Extend

1 Draw a distance–time graph for the following bike ride. Use kilometres and hours as the units for distance and time.

• The bike travels 10 km in the first hour.

• It is then stationary for two hours.

• It then travels 40 km in the next two hours.

• It is then stationary for one hour.

Use the graph to calculate the bike's speed during each part of the journey, and then the average speed for the whole journey.

≫ Relative speed

Worked example

A train is travelling at 50 km/h. A car on a road running parallel next to the track is travelling at 30 km/h.

Calculate the relative speed of the train and car when:

a) the car and train are travelling in the same direction
b) the car and train are travelling in opposite directions.

a) When two objects are travelling in the same direction, the relative speed between them is the difference between their speeds.

relative speed = fastest speed − slowest speed
relative speed = 50 km/h − 30 km/h = 20 km/h

b) When two objects are travelling in opposite directions, the relative speed between them is the sum of their speeds.

relative speed = fastest speed + slowest speed
relative speed = 50 km/h + 30 km/h = 80 km/h

Apply

1   Abdul and Rachel start at the same point. Abdul walks north at 3 km/h and Rachel walks south at 5 km/h.

   a) How far does Abdul travel in three hours?

   b) How long does it take Rachel to travel this far?

   c) What is the relative speed between Abdul and Rachel?

   d) How far apart are Abdul and Rachel after four hours?

   e) How far apart are Abdul and Rachel after Rachel has travelled 15 km?

Extend

1 Two trains start at the same station but travel in opposite directions. Train A travels at 45 km/h and train B travels at 70 km/h.

   a) What is the relative speed of the two trains?

   b) How far apart are the two trains after 2 hours and 45 minutes?

   c) How long does it take for the trains to be 500 km apart?

2 Passengers on a train often feel like they are moving backwards compared with the train next to them when they are in fact stationary. Explain why this happens with reference to relative motion.

» Acceleration

Worked example

A Formula 1 racing car accelerates from stationary to 120 m/s in 4 s. A normal car reaches the same speed in 8 s. Which has a higher acceleration?

Acceleration tells us how quickly an object can speed up or slow down. Therefore, an object that reaches a speed in a shorter time than another has a higher rate of acceleration. As the Formula 1 car can reach 120 m/s in half the speed of the normal car, it must have the higher acceleration.

Apply

1 How do we represent acceleration on a distance–time graph?

2 Sketch a distance–time graph for objects A and B, where object B is speeding up more quickly than object A.

Extend

All falling objects have the same acceleration when they fall, if we ignore air resistance. Imagine that a ping-pong ball is dropped from the top of a multi-storey building. It takes 10 seconds to reach the ground, ignoring air resistance.

1 Would it take a bowling ball more time, less time or the same amount of time to fall the same distance? Explain why.

2 What would be the difference if we did not ignore air resistance? Explain why.

3 Acceleration is always caused by a force. What force causes falling objects to accelerate?

4 Knowing this, do you think an object falling on the Moon would accelerate more quickly or more slowly than one on the Earth? Why?

2 Gravity

≫ Mass and weight

Worked example

The suits worn by Neil Armstrong and other astronauts on the Apollo missions to the Moon had a mass of approximately 90 kg.

a) What was the weight of these suits at the Earth's surface?
b) On the Moon, the suits only weighed 144 N, because the force of gravity is weaker there. Calculate the gravitational field strength on the surface of the Moon.
c) Why is gravitational field strength weaker on the Moon than on Earth?

a) Using the equation

weight = mass × gravitational field strength

weight = 90 kg × 10 N/kg = 900 N

b) Here, we need to rearrange the equation so that:

gravitational field strength = $\dfrac{\text{weight}}{\text{mass}}$ = $\dfrac{144}{90}$ = 1.6 N/kg

c) The gravitational field strength of a planet or moon depends on its mass. The Moon is much smaller than the Earth, and so has less mass. Therefore, its gravitational field strength is smaller too.

> **Hint**
>
> Gravitational field strength (g) has a value of 10 N/kg at the Earth's surface.

Apply

1 'I weigh 60 kg.' What is wrong with this statement? How could it be corrected?

2 Put these objects in order of the gravitational field strength they would have at their surface then explain your thinking.

Sun	Moon	Earth	Jupiter

3 What is the weight of an 8 kg dog?

4 What is the mass of a bag of potatoes that has a weight of 50 N?

5 How would the gravitational field strength of Earth change if the Earth were

  a) more massive

  b) less massive?

6 A 2 kg mass weighs 46.2 N on Jupiter. What is the value of g on Jupiter?

> **Hint**
>
> When working out the weight of an object, the mass must be in kilograms.

Extend

1 Explain the difference between mass and weight as if you were teaching it to a five-year-old child.

2 Gravitational field strength is different on different planets.

a) Copy and complete the table below.

Planet	Mass of object (kg)	Weight of object on planet (N)	Gravitational field strength of planet (N/kg)
A	5	45	
B	100	3500	
C	40	440	
D	1	3.8	
E	2500	8000	

b) Which planets must be the least and most massive? How do you know?

c) How much would a 500 g bag of flour weigh on planets B and E?

d) A 3 kg cat weighs 33 N. What planet must the cat be on?

3 Tim Peake is the first British European Space Agency astronaut. What would happen to the mass and weight of Tim when he is in his astronaut suit if he

a) went to the Moon

b) put lead weights in his boots

c) went to Jupiter

d) went on a (successful!) diet

e) went deep into outer space?

4 If you stood on a set of scales, your mass (in kg) at the top of the Empire State Building in New York would seem to be less than at the bottom. Explain why.

» Comparing gravity and other forces

Apply

1 What is the difference between a contact force and a non-contact force? Give some examples of each.

2 State a similarity and a difference between gravity and magnetism.

3 The average apple weighs about 1 N. When an apple is hanging from a tree, this weight is balanced by another force. Draw a force diagram to show this.

Extend

1 Satellites can orbit at different distances around the Earth. Satellite A is orbiting in a low polar orbit, while satellite B is in an orbit over the equator.

   a) Draw a diagram to show the difference between these orbits.

   b) Which satellite will experience the largest force of attraction due to the gravity of the Earth?

   c) Therefore, which satellite do you think is travelling the fastest? Explain your reasoning.

   d) Both satellite A and B have the same mass. Will they both have the same weight or will one weigh more than the other? Why?

   e) Satellite B slows down. Predict what you think will happen to it.

3 Voltage and resistance

» Potential differences

Worked example

The circuit on the right contains three 3V cells and two identical bulbs.

a) What is the total potential difference provided by the cells?
b) What would the potential difference across each bulb be?
c) One of the bulbs is taken away. What would the potential difference across the remaining bulb be now?
d) One of the cells is taken away. What would the potential difference across the single bulb be in this situation?

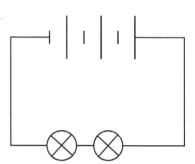

a) The three cells are all connected in the same direction, so the potential difference across them is simply 3 × 3V = 9V.

b) The bulbs are identical and the circuit is connected in series. This means that the potential difference provided by the cells must be shared equally between the two bulbs. Therefore, each bulb has

$\frac{9V}{2}$ = 4.5V of potential difference across it.

c) The total potential difference across all components in a series circuit must equal the potential difference across the power supply. As there is now only one component – the single bulb – the potential difference across this must be 9V.

d) Taking a cell away reduces the potential difference of the power supply to 2 × 3V = 6V. Therefore, the potential difference across the single bulb must be 6V too, as the bulb cannot have a higher voltage than the power supply.

> **Hint**
>
> A component is a device in a circuit that uses electrical energy and transfers it to some other form. For example, a light bulb transfers electrical energy to light (and heat) energy.

> **Hint**
>
> If a cell is connected in the opposite direction to the other cell, its potential difference must be taken away to find the total potential difference across the battery.

Apply

1 Each of the cells in the diagrams below has a potential difference of 1.5V. Calculate the potential difference across the following batteries of cells.

a)    b)    c)

2 The bulbs and cells in the diagram below are all identical. How would the brightness of the bulbs in circuits B, C and D compare with that of the bulb in circuit A? Explain why.

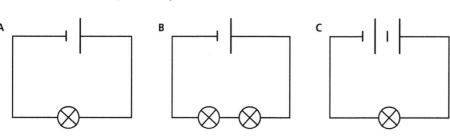 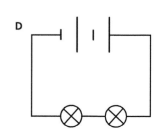

3 What must the potential difference be across the cells in circuits a)–c)?

a)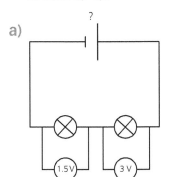

b)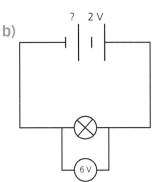

c)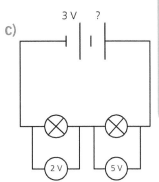

> **Hint**
>
> Identical components have the same potential difference across them in a series circuit.

Extend

1 Bulb A has twice the potential difference across it as bulb B. How would it look different? Why would this be the case?

2 Three identical bulbs are put in series with a 12V cell.

    a) Draw a circuit diagram for this.

    b) What must the potential difference across each bulb be?

    c) How would this change if a fourth bulb were added in series?

    d) How would this be different if only two bulbs were in the circuit?

3 Each of the cells in the two circuits below has a potential difference of 2V.

A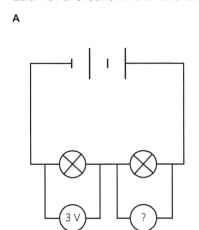

B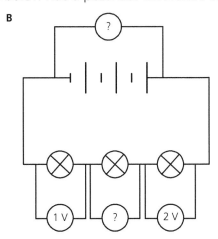

    a) Fill in the missing values on the voltmeters.

    b) Look at circuit B. Predict which bulb will be the brightest and why.

    c) Look at circuit A. What do you think the voltages will be across the two bulbs if a third 2V cell is added?

» Resistance

Worked example

The circuit on the right contains a cell and two resistors (A and B) in series. The current at all points is 2 A.

a) What must the potential difference across resistor B be?

b) Calculate the resistance of the two resistors.

c) What is the total resistance of the circuit?

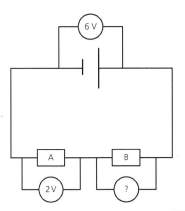

a) The potential difference across the cell is shared between components in a series circuit. Therefore the potential difference across resistor B must be $6V - 2V = 4V$.

b) For resistor A:

$$\text{resistance} = \frac{\text{voltage}}{\text{current}} = \frac{2V}{2A} = 1\Omega$$

For resistor B:

$$\text{resistance} = \frac{\text{voltage}}{\text{current}} = \frac{4V}{2A} = 2\Omega$$

c) Resistors in series add up. Therefore the total resistance $= 1\Omega + 2\Omega = 3\Omega$.

Apply

1 A component is labelled 'ohmic'. What does this mean?

2 Which would have a higher resistance, iron or wood? Why?

3 A resistor has a potential difference across it of 5V and current passing through it of 2A. Calculate its resistance.

4 What would the potential difference across a $10\,\Omega$ resistor be when a current of 3A passes through it?

5 Calculate the missing values in the resistor networks below.

a)
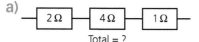
Total = ?

b)

Total = 8 Ω

c)
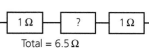
Total = 6.5 Ω

6 Look at the circuit below. The total resistance of the circuit is $10\,\Omega$.

   a) What must the resistance of resistor C be?

   b) What must the potential difference across resistor C be?

   c) Calculate the current passing through each of the resistors. What do you notice?

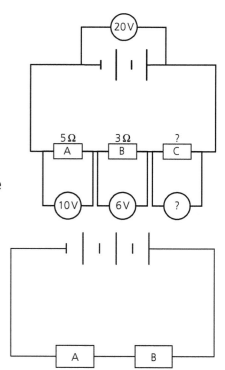

Extend

1 Look at the circuit on the right. Each cell has a potential difference of 2V.

   a) What must the potential difference across the battery of cells be?

   b) What must the potential difference across resistors A and B together be?

   c) The current running through the whole circuit is 4A. Using this, calculate the total resistance of the circuit.

   d) If resistor A has half the resistance of resistor B, what must the resistances of resistors A and B be?

4 Current

» Multiple loops

Worked example

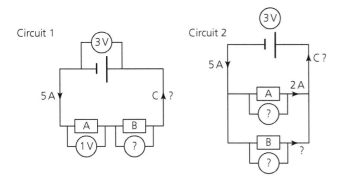

These two circuits both contain a 3V cell and two resistors. Circuit 1 is connected in series and circuit 2 is connected in parallel.

For circuit 1:

a) What would the potential difference across resistor B be?
b) What would the current at point C be?

For circuit 2:

c) What would the potential difference across resistors A and B be?
d) What would the current through resistor B and at point C be?

For circuit 1:

a) In a series circuit, the potential difference across all of the components in total must equal the potential difference across the power supply. Therefore, the potential difference across B must equal 3V – 1V = 2V.

b) In a series circuit, the current is the same everywhere. Therefore, the current at point C must also be 5A.

For circuit 2:

c) In a parallel circuit, the potential difference across each branch is equal to the potential difference across the power supply. Therefore, the potential difference across resistors A and B must each be 3V.

d) In a parallel circuit, the current through the branches must add up to the current passing through the power supply. Therefore, the current through resistor B must be 5A – 2A = 3A. The current at point C must be 5A, as this is when the branches have joined back together again.

> **Hint**
>
> The potential difference across or the current running through a resistor is not fixed – it can change, depending on the arrangement of the circuit that it is in.

Know

1 Which of the following are series circuits and which are parallel?

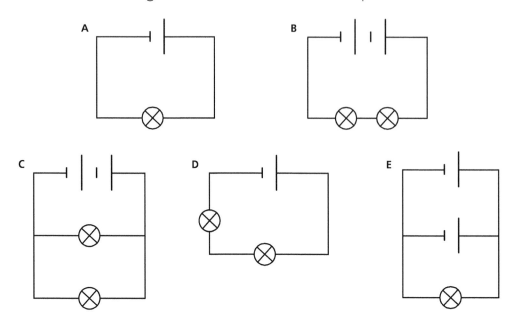

Apply

1 A bulb is connected to a battery. What will happen to the brightness of this bulb if an identical bulb is connected

   a) in series

   b) in parallel?

   Explain why.

2 Look at the circuit on the right. Three identical bulbs are connected with a power supply in parallel.

   a) If the current running through bulb A is 2A, what current must be passing through bulbs B and C?

   b) What would the current passing through the battery be?

   c) If the resistance of each bulb is 5Ω, what must the potential difference across each bulb be?

   d) Therefore, what must the potential difference across the battery be?

   e) Therefore, what must the potential difference across each of the two identical cells be individually?

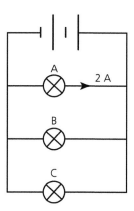

3 A 10Ω and 20Ω resistor are connected to a 6V battery in

   a) series

   b) parallel.

   For each of these situations, draw a circuit diagram of the set up and determine the potential difference across each resistor.

Extend

1 The circuit on the right contains a cell and two resistors in parallel.

   a) What must the current passing through the cell be?

   b) What is the potential difference across resistors A and B?

   c) Calculate the resistance of resistors A and B.

2 In old sets of fairy lights, the bulbs were all connected to the power supply in series. Nowadays, they are connected in parallel. Explain why this is an advantage.

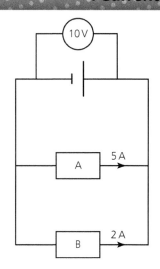

» Static charge

Know

1 Will the following charges repel or attract?

   a)   b)   c)   d)   e)   f)

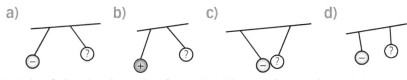

Apply

1 What is created if positive and negative charges are separated?

2 What must the charges be in the following pairs?

   a)   b)   c)   d)

3 Put the following in order, from the biggest force of repulsion to the biggest force of attraction.

   A   B   C   D   E

4 Static charge can be created by rubbing a balloon on a jumper or on your hair.

   a) Describe how the balloon becomes charged.

   b) A negative balloon sticks to the wall. What charge must the wall have? Why?

Extend

When you touch a Van de Graff generator, you become statically charged; electrons are transferred to or away from you. This can cause your hair to stand on end.

1 What way must the electrons have been transferred if you become

   a) negatively charged

   b) positively charged?

2 Are your strands of hair being attracted to or repulsed from each other? Why?

3 Predict what would happen if you kept hold of the Van de Graff generator for longer.

4 When using a Van de Graff generator, you should stand on plastic or wood.

a) Why should you stand on plastic or wood?

b) Predict what would happen if you stood on metal. Why would this happen?

» Models

Extend

A common model used for electric circuits involves water pipes, boilers and radiators. Look at the two diagrams below, representing series and parallel circuits in this model.

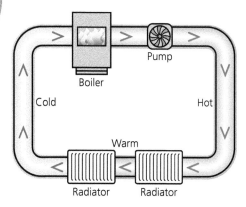
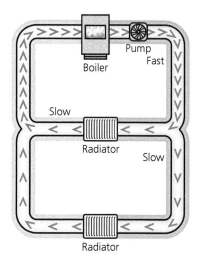

1 What represents the following in this model:

a) battery

b) components

c) current

d) potential difference

e) resistance

f) a series circuit

g) a parallel circuit?

2 Evaluate this model by giving two strengths and two weaknesses.

3 Can you come up with another model for electric circuits to help you better understand how they work?

5 Energy costs

» Energy

Worked example

A cyclist transfers about 800 kJ of energy every hour.

a) How many kilojoules (kJ) of energy will a cyclist transfer in half an hour?
b) 100 g of chocolate contains about 2000 kJ. How long would it take a cyclist to transfer this much energy?

a) If the cyclist transfers 800 kJ in one hour, he or she will transfer 800 × 0.5 = 400 kJ in half an hour.

b) For this, we need to divide the energy in 100 g of chocolate by the energy a cyclist transfers per hour. The time would equal

$$\frac{2000 \text{ kJ}}{800 \text{ kJ}} = 2.5 \text{ h}$$

Apply

1 Different types of exercise 'burn off' different amounts of energy. The table shows how much energy various activities transfer each half an hour.

Activity	Energy transferred per half an hour, in kJ
Aerobics	550
Basketball	700
Bowling	250
Dancing	450
Jogging	800
Stair climbing	650
Swimming	500
Walking	350

   a) Which activity transfers the most energy per half an hour?

   b) How much energy will you transfer if you

      i) play basketball for one hour   iii) jog for 15 minutes

      ii) bowl for two hours   iv) walk for four hours?

   c) Which would need more energy – dancing for four hours or doing aerobics for three hours?

   d) How many more kilojoules of energy would you transfer if you spent 15 minutes stair climbing rather than swimming?

2 Carbohydrates contain 16 kJ of energy per gram, while fat contains about 37 kJ per gram. A digestive biscuit contains 6 g of fat and 19 g of carbohydrates. What is the total energy provided by the carbohydrate and fat in one biscuit?

» Energy resources

Know

1 Sort the following energy resources into 'renewable' and 'non-renewable' and then give an advantage and a disadvantage for each:

   a) coal   d) nuclear fuel   g) waves

   b) solar   e) oil   h) gas

   c) wind   f) biomass   i) geothermal.

Apply

1 Make a flow chart describing how fossil fuels were formed.

2 The world currently makes most of its energy using non-renewable energy sources. Suggest reasons why we are starting to use more renewable energy sources to make electricity.

3 Two countries (A and B) use the following proportions of energy resources.

a) Which country uses a higher percentage of fossil fuels? How much more?

b) Which country uses a higher percentage of renewable energy? How much more?

c) Which country is more likely to be nearer the equator? Why?

d) Which country is more likely to be an island? Why?

e) In total, country A transfers 200 MJ of energy each day, while country B transfers 500 MJ. How many megajoules (MJ) must be produced by coal and wind in each country?

Energy source	Percentage used by country A	B
Coal	40%	30%
Oil	10%	30%
Gas	20%	20%
Nuclear	5%	0%
Wind	10%	1%
Solar	2%	15%
Tidal	8%	2%
Geothermal	5%	2%

Extend

1 Even though fossil fuels are running out and damage the environment, most countries still use them for most of their energy needs. Suggest some reasons why.

2 Which energy sources would be the most suitable for the following countries/states? Decide on two energy sources for each country/state and explain your reasoning. You may have to do some research to find out the conditions in each location.

a) UK   c) Jamaica   e) Iceland

b) Egypt   d) Canada

» Energy in the home

Worked example

If electricity costs 11 p per kilowatt-hour (kWh), how much does it cost to operate a 2000 W kettle for three hours and 15 minutes?

First, we need to convert 2000 W into kilowatts. As 1000 W = 1 kW, 2000 W must be 2 kW.
We also need to convert time into hours. Fifteen minutes is a quarter of an hour, so three hours and 15 minutes must be 3.25 hours.
We can now put these into the equation.

cost = power (kW) × time (hours) × price per kWh

cost = 2 kW × 3.25 h × 11 p = 71.5 p

Hint

To answer this question, we need to use the equation

cost = power (kW) × time (hours) × price per kWh

It is important to have time in hours and power in kilowatts (kW). 1 kW equals 1000 W.

Apply

1 Complete the table below. The first row has been done for you.

Cost (p)	Power (kW)	Time (h)	Cost per kWh (p)
110	5	2	11
	3	6	11
	2	4	14
	1.5	7	14
80	8	1	
100		5	10
40	2		5

2 Here are some typical power ratings for everyday appliances. Fill in the gaps.

Appliance	Power (W)	Power (kW)
Light bulb	100	0.1
Desktop computer	300	
Kettle		2
Fridge		0.2
Hairdryer	1500	
Electric shower		8
Washing machine	500	

3 Using the table in Question 2 above, and assuming that electricity costs 11 p per kilowatt-hour (kWh), how much would it cost to use

a) the fridge all day

b) the washing machine for two hours

c) a hairdryer for 15 minutes

d) an electric shower for 20 minutes

e) five light bulbs for six hours

f) a desktop computer for three hours and a kettle for half an hour?

Extend

1 Using the table in Question 2 of the Apply section above, answer the following questions.

a) Which would cost more – using the desktop computer for two hours or the hairdryer for half an hour?

b) How much does it cost to run the fridge for a week?

c) How long could you run the kettle for on £1?

6 Energy transfer

» Energy transfers

Worked example

A pop-up toy is pressed down and then released. Draw simple energy transfer diagrams for this situation.

When the toy is pressed down, it fills its elastic store of energy. When the toy is released, this energy is transferred to its gravitational store, as the toy rises up in the air.

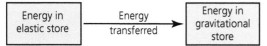

| Energy in elastic store | —Energy transferred→ | Energy in gravitational store |

If we want to go one step further, we can think of the next energy transfer. Once the toy has reached the peak of its journey, it will fall back down again, transferring energy to its kinetic store as it speeds up.

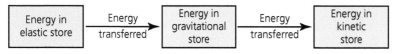

| Energy in elastic store | —Energy transferred→ | Energy in gravitational store | —Energy transferred→ | Energy in kinetic store |

Apply

1 Draw simple energy transfer diagrams for the following situations:

a) dropping an apple from a height

b) heating up soup in a pan on a gas hob

c) throwing a ball into the air

d) jumping on a trampoline

e) powering a motor with a battery.

Extend

1 Rollercoasters use energy transfers to work. Look at the rollercoaster diagram below.

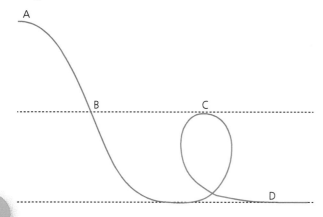

a) What stores of energy will be filled at points A and D?

b) What stores will be empty at points A and D?

c) Where will the carriage have equal stores of kinetic and gravitational energy?

d) If the carriage has 1000 J of energy in its gravitational store at point A, work out how much energy it will have in its gravitational and kinetic stores at points A–D.

e) So far in this question, we have ignored friction. If we now consider friction, what effect will this have on the rollercoaster's energy stores throughout its journey.

» Energy dissipation

Worked example

A battery provides 400 J of energy for an electric buzzer. However, the buzzer only transfers 250 J of this energy.

a) What has happened to the energy that is 'missing'?
b) Which energy store has the 'missing' energy been transferred to?
c) How much energy has been transferred in this way?

a) The 'missing' energy has not disappeared – it has been dissipated as heat. This means that it has been spread out and shared between more energy stores.
b) The energy has been transferred to the thermal stores of the surroundings – e.g. the wires of the circuit, the buzzer itself and the air around it.
c) Energy cannot be created or destroyed, so the energy dissipated must be equal to the difference between the energy provided by the battery and the energy used by the buzzer. Therefore, 400 J – 250 J = 150 J has been dissipated.

Apply

1 A ball is dropped from a height. It originally had 25 J of energy in its gravitational store.

a) How much energy will it have in its kinetic store just before it hits the ground, if no energy has been dissipated?

b) In fact, the ball only has 22 J of energy in its kinetic store at this point.

How much energy has been dissipated?

c) What percentage of energy has been wasted?

d) What energy stores have been increased as a result of this dissipation?

e) How does this affect the speed of the ball?

Extend

1 Parachutes work by purposefully dissipating energy. Discuss how the energy transfers and energy dissipation are different for a skydiver wearing a large parachute compared with one wearing a smaller parachute. How does this affect their landing speed? Why?

2 A lorry dissipates a higher percentage of its energy than a sports car does.

a) Discuss why you think this is.

b) Suggest methods of reducing the amount of energy dissipation for either vehicle.

» Consequences of energy dissipation

Apply

1 What is a perpetual motion machine? Explain why it is impossible to build one.

2 A pendulum in a grandfather clock swings from side to side to keep time.

a) Where is the pendulum stationary?

b) Where is the pendulum travelling the fastest?

c) What energy stores are filled at points A, B and C?

d) Draw an energy transfer diagram to explain which energy stores are being filled at which time.

e) In an ideal world, the pendulum would swing forever without slowing down or stopping. Explain why this is not the case in the real world.

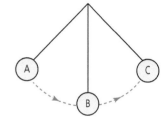

Extend

1 Here is a table of the efficiencies of various light bulbs. The efficiency of an appliance tells us what percentage of the energy transferred into a system is used usefully.

Bulb	Efficiency
A	10%
B	70%
C	50%
D	20%

a) Which bulb is the most efficient?

b) Which bulb wastes the most energy?

c) What energy store would this waste energy be dissipated into?

d) If each bulb uses 60J of energy per second, determine how much energy each bulb dissipates.

e) If bulb B uses 80J of energy per second and bulb C uses 110J of energy per second, which bulb wastes more energy per second?

f) How much energy must be transferred to bulb B per second if it usefully transfers 10J per second as light?

7 Sound

» Describing sound

Know

1 Put these states of matter in order of how fast sound travels through them, from slowest to fastest: **solid**, **liquid**, **gas**.

2 Define the following key words and then label a) and b) on a copy of the diagram:

   a) amplitude

   b) period

   c) frequency.

Apply

1 Sound is a longitudinal wave. What does this mean?

2 Why can sound not travel through a vacuum?

3 A bell rings. Describe how the sound travels from the bell to your ear.

4 What's wrong with the following wave diagrams?

a)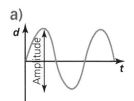

b)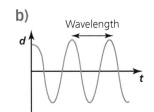

c)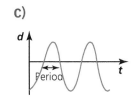

5 Look at the wave diagrams below.

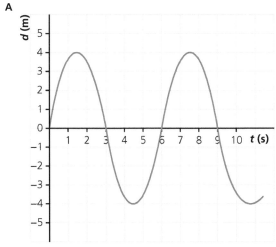

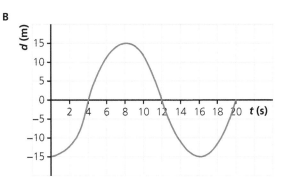

   a) Measure the period and amplitude of each wave.

   b) Which would be the loudest? Why?

   c) Which would have the lowest pitch? Why?

6 Two hundred waves pass a point in five seconds. What is the frequency?

7 What is the frequency of a wave where 120 waves pass a point per minute?

8 Sketch

   a) a quiet, high-pitched sound wave

   b) a loud, low-pitched sound wave

   c) a low-frequency, small-amplitude sound wave

   d) a high-frequency, large-amplitude sound wave.

Extend

1 Draw the following waves on squared paper.

Wave	Amplitude	Period
A	3 cm	5 s
B	2 cm	2 s
C	4 cm	4 s
D	6 cm	10 s

2 Of the waves in the table in Question 1 above, explain which would be the

   a) loudest? Why?   **c)** highest pitched? Why?

   b) quietest? Why?   **d)** lowest pitched? Why?

3 Look at the four wave diagrams below.

Put them in order from

   a) lowest to highest frequency

   b) largest to smallest amplitude

   c) shortest to longest period.

» Ranges

Know

1 What is the average hearing range for humans?

2 What is ultrasound?

3 How many hertz are in 1 kHz?

4 What is an echo? Give two examples of echoes in everyday life.

Apply

1 State some uses of ultrasound.

2 Look at the diagram of the ear on the right.

   a) Label a copy of the diagram using the key words ear canal, eardrum, bones and cochlea.

   b) Give a simple description of the function of each of these parts.

   c) Give some examples of how human hearing can be damaged.

Extend

1 Explain how sound is transmitted through the ear, including the cochlea.

2 Sonar is used by ships and other vessels to measure the depth of the ocean and look for objects in the sea. Using the diagram, explain how you think sonar works.

3 Research why pregnant women are given ultrasound scans to see the baby inside their body, rather than X-rays.

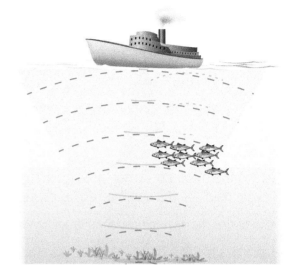

≫ Insulation

Know

1 What is noise pollution?

2 Give some everyday examples of noise pollution.

3 How does insulation reduce noise pollution?

4 Give some examples of how you can insulate your house against noise pollution.

Apply

1 Why are decibels (dB) not measured on a linear scale?

2 How much louder is a 40 dB sound than a 10 dB sound?

3 Why would triple glazing reduce noise pollution more than double glazing?

Extend

1 Workers on construction sites have to wear ear protectors to insulate their ears and reduce noise pollution. Design your own pair of ear protectors, explaining the choices you have made.

8 Light

» Transmitted light

Know

1 What happens when light hits an object and is

a) absorbed

c) reflected

b) transmitted

d) refracted?

2 What is the law of reflection?

3 What is refraction? When does it happen?

Apply

1 Copy the diagram on the right and add two light rays to show how the eye would see the apple reflected in the mirror.

2 Copy the diagram below then label it with the following key words

a) glass block

d) refracted ray

b) normal (×2)

e) angle of incidence (×2)

c) incident ray

f) angle of refraction (×2).

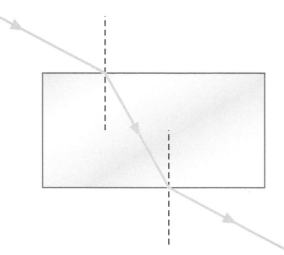

3 Copy and complete the following refraction diagrams.

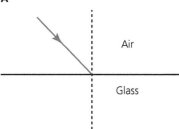

A

Air

Glass

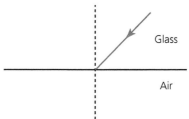

B

Glass

Air

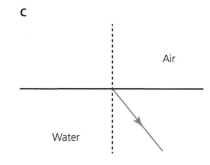

C

Air

Water

Extend

1 Periscopes are used in submarines to see what is happening on the surface of the sea. Research how they use mirrors to do this and draw a ray diagram to help you explain.

2 Refraction happens due to a change in wave speed for the light. How do you think the following would affect the speed of light? Why?

   a) Light entering glass from air.

   b) Light leaving water into air.

3 Glass is denser than water. Draw a ray diagram showing how light would refract if it went

   a) from water to glass

   b) from glass to water.

4 Putting a straw in water causes it to appear bent. Can you explain why?

» Colours

Worked example

White light travels through two filters; the first is magenta and the second is red.

a) What colour(s) of light are transmitted through this combination?
b) What colour would a white T-shirt, red jumper and blue pair of jeans look through these filters?

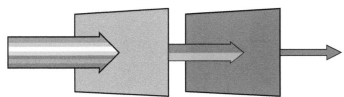

> **Hint**
>
> A white object reflects all colours of light. A black object reflects none. A coloured object (e.g. red), reflects only that colour and absorbs the rest.

a) The magenta filter absorbs all light that is not blue or red. The red filter then absorbs the blue light and only allows the red light to be transmitted.
b) Under the remaining red light, a white T-shirt would look red, as would the red jumper. The blue jeans would look black, as blue objects absorb all colours of light except blue, which they reflect.

Know

1 Copy and complete this diagram by filling in the colours of the spectrum.

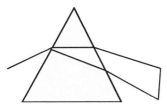

2 What are the three primary colours of light?

Apply

1 Copy and complete the colour chart on the right to show how light can be combined to make different colours.

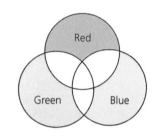

2 Copy and complete the table below to show which colours of light the filters will let through.

Filter colour	Will the filter transmit:		
	blue light?	red light?	green light?
Blue			
Red			
Green			
Yellow			
Cyan			
Magenta			

Extend

1 White light travels through the following filters:

   a) blue then red

   b) yellow then green

   c) cyan then magenta.

   What colour(s) of light (if any) are transmitted through each combination?

2 What colours will flowers 1–3 look when viewed through the following filters:

   a) red   **c)** blue   **e)** magenta

   b) green   **d)** cyan   **f)** yellow?

Flower 1   Flower 2   Flower 3

9 Particle model

» Model of solids, liquids and gases

Worked example

Explain, using the particle model, why a solid bar of iron expands when it is heated.

The iron particles gain energy and vibrate more vigorously. As a result the particles move further apart.

Hint

When a solid is heated, its particles do not get bigger. They stay the same size. Stating that the particles get bigger is a very common mistake.

Know

1 How are the particles arranged in:

   a) a solid   b) a liquid   c) a gas?

2 Describe how the movement of the water particles differs between ice, water and water vapour.

3 The diagram shows the three states of matter for a substance.

Each circle represents a particle of the substance.

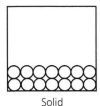 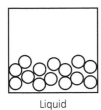 

Solid   Liquid   Gas

   a) Copy and complete the diagram by drawing three circles to represent the particles of a gas.

   b) Which statement is correct about the movement or arrangement of the particles of this substance?

   A They move randomly in the liquid state.

   B They move randomly in the solid state.

   C They are arranged in fixed positions in the gas state.

   D They are arranged in fixed positions in the liquid state.

   c) Which word describes the change that takes place when a solid becomes a liquid?

   A boiling   C freezing

   B condensing   D melting

Apply

1 The table below gives the melting and boiling points of five substances: A, B, C, D and E.

Substance	Melting point (°C)	Boiling point (°C)
A	−270	−269
B	600 (sublimes)	×
C	770	1420
D	−7	59
E	0	100

a) State the physical state of each substance at 20 °C.

b) State and explain which substance has the strongest attractions between its particles.

c) State and explain which substance has the weakest attractions between its particles.

d) State and explain which substance has the greatest distance between its particles at 20 °C.

e) Why is no boiling point given for substance B?

Extend

1 In 1827, Robert Brown used a microscope to view pollen grains suspended in water.

He noticed that the pollen grains were moving around randomly. However, he could not see anything hitting the pollen grains.

This movement was subsequently called Brownian motion.

Suggest how Brownian motion provides evidence that water is made up of tiny particles.

» Changes of state

Worked example

Iron has a melting point of 1535 °C and a boiling point of 3000 °C. Work out what state iron is in at each of the following temperatures: 1600 °C, 200 °C, 3200 °C.

At 1600 °C the iron is above its melting point but below its boiling point, so it is a liquid.
At 200 °C the iron is below its melting point, so it is a solid.
At 3200 °C the iron is above its boiling point, so it is a gas.

Know

1 What is the name of the process that occurs when a liquid changes into a gas at room temperature?

    A condensation

    B diffusion

    C evaporation

    D sublimation

2 The table below shows the melting and boiling points of four elements, A, B, C and D.

Element	Melting point (°C)	Boiling point (°C)
A	660	2520
B	1540	2760
C	650	1100
D	−39	357

Which element in the table is:

a) a liquid at 0 °C

b) a solid at 1500 °C

c) a gas at 500 °C

d) a liquid over the biggest temperature range?

Apply

1 When a liquid evaporates at room temperature, it changes into a gas.

The diagram shows the arrangement of the particles in a liquid.

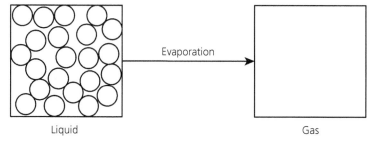

Liquid   Evaporation   Gas

a) Copy the box for the gas and show the arrangement of three particles in the gas.

b) Describe the movement of the particles in a gas.

c) Use the particle model to explain why heating a liquid causes it to evaporate more quickly.

Extend

1 The apparatus in the diagram is used to measure the melting point of solid Y.

The liquid is gently heated and the temperature at which solid Y melts is recorded.

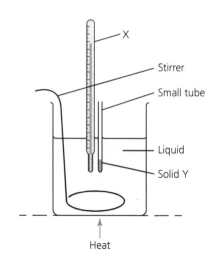

a) Give the name of the piece of apparatus labelled X.

b) Solid Y melts at 150 °C. Explain why water is not a suitable liquid to use in this experiment.

c) Suggest why the liquid in the beaker needs to be stirred constantly.

» Sublimation, diffusion and pressure

Worked example

a) Use the particle model to explain how air inside a balloon creates pressure on the sides of the balloon.

b) Why does the pressure increase if you blow more air into the balloon?

a) The air particles are continually colliding with each other and with the sides of the balloon. When a particle collides with the wall, it exerts a small force on the wall. The pressure exerted by the air is a result of the collision forces of all of the particles.

b) When you blow more air into the balloon, you add more particles. The more particles that hit the walls, the higher the pressure.

Know

1 What is the name of the process that occurs when a solid changes straight into a gas?

A condensation   C evaporation

B diffusion   D sublimation

2 What is the name of the process when the gas spreads out to fill the gas jar?

A condensation   C evaporation

B diffusion   D sublimation

Apply

1 Compressed air from a can is used to clean computer keyboards.

a) Use the particle model to explain how a gas causes a pressure on the inside of the container.

b) The can has a warning sign.

How would increasing the temperature of the compressed air affect the pressure in the can? Explain your answer.

Extend

1 When ammonia gas and hydrogen chloride gas mix, they react to form a white solid.

A cotton wool pad is soaked in ammonia solution and another is soaked in hydrogen chloride solution. The two pads are then put into opposite ends of a dry glass tube at the same time.

After five minutes, a white solid forms as shown in the diagram below.

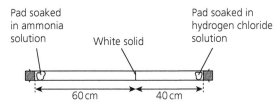

Pad soaked in ammonia solution   White solid   Pad soaked in hydrogen chloride solution

60 cm   40 cm

a) Describe, in terms of particles, what happens in order for the white solid to form.

b) Suggest why the white solid does not form in the middle of the tube.

c) The experiment is repeated at a higher temperature. State and explain how this change would affect the time taken for the white solid to form.

d) Gas particles move at a speed of several hundred metres per second at room temperature. Suggest why it took five minutes for the white solid to form.

2 A student plays a trick on his classmates. He lets off a stink bomb at the front of the class without telling them. He measures how long it takes for different students to notice.

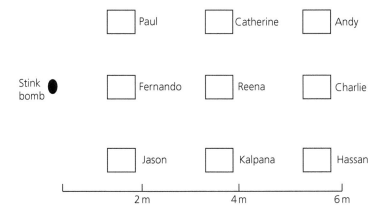

Paul   Catherine   Andy

Stink bomb   Fernando   Reena   Charlie

Jason   Kalpana   Hassan

2 m   4 m   6 m

The table shows his results.

Name	Paul	Fernando	Jason	Catherine	Reena	Kalpana	Andy	Charlie	Hassan
Time in s	13	10	13	24	20	24	35		35

a) Explain how the 'smell particles' reach the students.

b) Why is Fernando the first to notice the smell?

c) Use the results to estimate how long it takes for Charlie to notice the smell.

d) How fast do the smell particles travel through the air? Calculate the speed in metres per second (m/s).

3 Gases do not all diffuse at the same speed. The speed at which they diffuse depends on the mass of their particles.

The apparatus shown in the diagram is used to investigate the speed at which gases diffuse.

The plunger is allowed to fall under its own weight, pushing the gas out of the syringe and through the tube containing the glass wool.

The time taken for 100 cm³ of each gas to escape is measured.

The graph shows the results for six different gases.

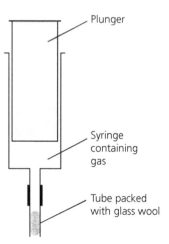

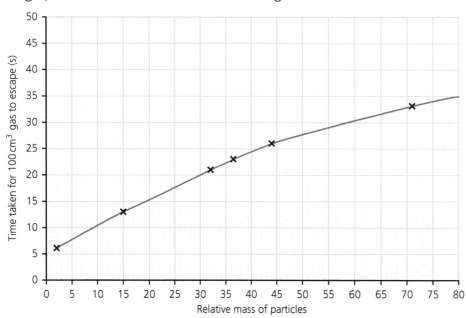

a) State the relationship between the time taken for the gas to escape and the relative mass of the particles.

b) A gas has particles with a relative mass of 65. Use the graph to find the time taken for 100 cm³ of this gas to escape from the apparatus.

c) 100 cm³ of air escapes from the apparatus in 20 s. Use the graph to find the average relative mass of a particle of air.

d) Suggest why a balloon filled with hydrogen gas gets smaller more quickly than a balloon filled with air at the same temperature and pressure.

10 Separating mixtures

» Pure substances and mixtures

Worked example

Hydrated copper sulfate is a blue solid that is soluble in water. Lead carbonate is a white solid that is insoluble in water. Describe and explain what you would see if a small sample of each of these solids was added separately to water and the mixture was stirred.

When the copper sulfate is added to water and stirred, the blue solid disappears and the water turns blue. The copper sulfate dissolves. The particles of copper sulfate have broken away from the solid and have spread evenly throughout the water.

When lead carbonate is added to water and stirred, the white solid does not disappear and the water goes cloudy white. The lead carbonate does not dissolve. The particles of the solid remain together.

> **Hint** ❗
>
> When answering a question that asks for both a description and an explanation, give the description first and then give the explanation in a separate sentence.

Know

1 When sugar is added to water and the mixture is stirred, the sugar disappears. A possible explanation for this is that the sugar _____ in water.

   Choose from the list a word that best fills the gap in the paragraph above.

   A condenses   B dissolves   C evaporates   D melts

2 State whether each of the following is a pure substance or a mixture of substances.

   a) air

   b) iron

   c) sodium chloride

   d) mineral water

   e) sea water

3 Some salt is added to water. The mixture is stirred and the salt dissolves. Choose a word from the box to complete the sentences that follow.

solute	solvent	solution

   a) The salt is the _____.

   b) The water is the _____.

   c) The mixture of salt and water is a _____.

Apply

1 Salol is a solid that melts at 42 °C to form liquid salol. Liquid salol and a solution of salol are both colourless liquids.

   a) What is the difference between liquid salol and a solution of salol?

   b) How would you obtain solid salol from a solution of salol?

   c) How would you obtain solid salol from liquid salol?

2 Tincture of iodine is a mixture of solid iodine dissolved in liquid alcohol. Give the name of:

   a) the solvent

   b) the solute

   in tincture of iodine.

Extend

1 A student wants to find out if chalk is soluble in water.

   This is the method he uses:

   (1) Add some pure powdered chalk to pure water.

   (2) Stir the mixture and leave it for a while.

   (3) Filter the mixture and collect the filtrate in an evaporating dish.

   (4) Leave the evaporating dish in a warm place until all of the water has evaporated.

   After evaporation, a small amount of a white solid is left in the evaporation dish.

   a) Why did the student stir the mixture of chalk and water and then leave it for a while?

   b) Why did the student filter the mixture of chalk and water?

   c) What deduction can be made from the result of the experiment?

2 Benzoic acid is a white solid that is insoluble in cold water, but soluble in hot water.

   Charcoal is a black solid that is insoluble in both cold and hot water.

   a) Explain how you could obtain pure benzoic acid from a mixture of benzoic acid and charcoal.

   b) How could you show that the benzoic acid is pure?

3 The graph below shows how the solubility of three salts – potassium chloride, sodium chloride and calcium sulfate – changes as the temperature changes.

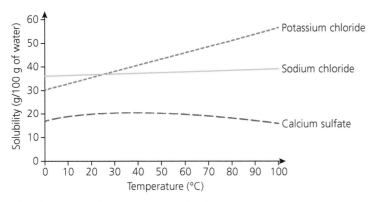

a) Which salt is the most soluble at

   i) 10 °C

   ii) 60 °C?

b) At what temperature do potassium chloride and sodium chloride have the same solubility?

c) A beaker contains 51 g of potassium chloride dissolved in 100 g of water at 80 °C.

State and explain what you would see if this solution were allowed to cool to 40 °C.

d) The water in a lake had the three salts dissolved in it. The water in the lake evaporated and the salts were deposited as solids in the order shown in the diagram.

   i) Use the graph to explain how you know that these salts were deposited when the temperature of the water was above 25 °C.

   ii) Which salt would be on top and which salt would be on the bottom if the salts were deposited when the temperature of the water was 10 °C?

» Purifying liquids

Worked example

Coffee can be made from ground coffee by placing the ground coffee in a paper filter and pouring hot water onto the coffee. The solution of coffee can then be collected in a jug placed below the filter.

a) Why is hot water used rather than cold water?
b) Why is a filter used?
c) Why is it not necessary to use a filter when you make instant coffee?

a) The coffee will dissolve more quickly in hot water than in cold water.
b) To remove the substances that do not dissolve in water.
c) Instant coffee contains only substances that are soluble in water. The insoluble substances have been removed.

Know

1 Copy and complete each of the following sentences using words from the box.

gas	liquid	solid	solution

   a) Filtration is used to separate an insoluble solid from a _____ .

   b) Evaporation is used to separate a soluble solid from a _____ .

   c) Distillation is used to separate a _____ from a solution.

2 Chalk does not dissolve in water, but sugar does. A mixture of chalk and sugar is added to water and the mixture is stirred. The mixture is then filtered using a filter funnel and paper.

   a) Which substance remains in the filter paper after filtration?

   b) What does the liquid that passes through the filter paper consist of?

   c) The liquid that passes through the filter paper is left in a warm place. After 2 days all that is left is a white solid. What is the white solid?

3 Salt is soluble in water. Sand is insoluble in water. This difference allows a mixture of salt and sand to be separated using the apparatus shown in the diagram.

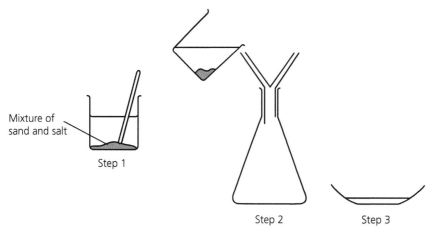

Mixture of sand and salt

Step 1

Step 2   Step 3

Use terms from the box to copy and complete the following sentences. Each term may be used once, more than once or not at all.

beaker	Bunsen burner	conical flask	
glass rod	thermometer	water	funnel

a) In Step 1, the mixture of salt and sand is put into a _____ containing _____ and the mixture is stirred with a _____.

b) In Step 2, the mixture is poured through a _____ into a _____.

c) In Step 3, the liquid from Step 2 is put into a basin to allow the _____ to evaporate.

Apply

1 Rock salt contains a mixture of sand, dirt and salt. Both sand and dirt are insoluble in water. Salt is soluble in water. Describe how you could separate and collect the salt from rock salt.

2 You have spilt some sugar into a packet of rice grains. Describe how you could obtain the rice from the mixture.

3 Four methods that can be used to separate substances in a mixture are:

• chromatography

• distillation

• evaporation

• filtration.

Choose the best method to:

a) obtain water from a solution of solid copper sulfate dissolved in water

b) obtain solid salt from a solution of salt in water

c) obtain solid chalk from a suspension of chalk in water

d) separate the coloured materials in a food dye.

Extend

1 A student wants to find the best design for an evaporating basin in order to obtain salt from a solution of salt in water.

She compares the three basins – A, B and C – shown below.

 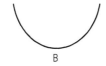 

A   B   C

a) Which basin would allow the water to evaporate most quickly? Give a reason for your answer.

b) Describe an experiment you could do to find out which basin is the best design. Include in your answer the steps you would take to make sure your experiment is a fair test.

2 Sand is insoluble in water and insoluble in tetrachloromethane (a colourless liquid).

Common salt is soluble in water, but insoluble in tetrachloromethane.

Wax is insoluble in water, but soluble in tetrachloromethane.

Explain how you could obtain a pure sample of each substance from a mixture of sand, common salt and wax.

3 The table below gives some information about three gases, X, Y and Z.

Gas	Solubility in water	Boiling point (°C)
X	Poor	10
Y	Poor	−210
Z	Good	−156

Use the information in the table to produce a method to separate the three gases using apparatus found in the laboratory.

4 A blue ink is made by dissolving a solid blue dye in water.

A student uses apparatus shown in the diagram to distil a sample of this blue ink.

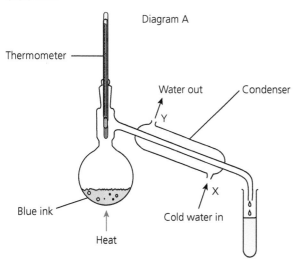

Diagram A

a) Which one of the following occurs during the distillation:

A condensation then evaporation

B evaporation then condensation

C melting then boiling

D melting then evaporation?

b) i) Name the colourless liquid that collects in the test tube.

ii) What is the reading on the thermometer as the colourless liquid distils over?

c) The cold water entering the condenser at X has a temperature of 20 °C.

Suggest a value for the temperature of the water leaving the condenser at Y.

Give a reason for your answer.

d) Another student uses the apparatus shown in diagram B to distil a separate sample of the blue ink.

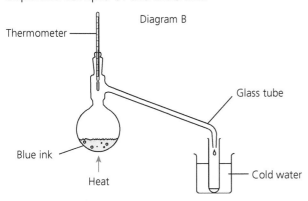

Diagram B

Explain why the condenser used in diagram A is better than the glass tube and beaker of cold water in diagram B?

» Chromatography

Worked example

A lime-flavoured ice-lolly is coloured green. State how you would show whether the ice-lolly contained a single green colouring or a mixture of yellow and blue colourings.

Allow the lolly to melt and then carry out chromatography on the liquid. If there is only a single green colouring present then there will be a single spot on the paper. If there is a mixture of yellow and blue, then there will be two spots on the paper. One will be coloured yellow, the other coloured blue.

Hint

The question does not ask for experimental details, so all that is necessary is to state the name of the method that should be used. Also, it is important to give the expected result if there is only a single colouring present, and if there is a mixture present.

Know

1 State what is meant by the term chromatography.

Apply

1 The diagram shows the result of an experiment to separate the colours present in four different dyes, P, Q, R and S.

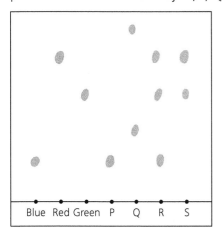

a) State and explain which dye contains three colours.

b) Each dye is made from one or more of the colours blue, red and green. State and explain which one of the results appears to be incorrect.

2 Universal indicator consists of a number of dyes dissolved in liquid ethanol.

a) What method could be used to separate the dyes?

b) What method could be used to obtain ethanol from the universal indicator?

c) How could you show that the ethanol obtained is pure?

Extend

1 The chromatogram below was obtained for a set of water-soluble felt-tip pens.

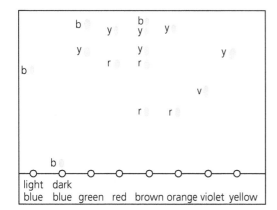

Colour of spots
b = blue
r = red
y = yellow
v = violet

light dark
blue blue green red brown orange violet yellow

a) Which pens contain only one dye?

b) i) Green can be made by mixing blue and yellow. Has the pen containing green ink been made this way?

   ii) Has the light blue dye, the dark blue dye or a different blue dye been used to make the green ink? Give your reasons.

c) Brown can be made from mixing red and green. How has the brown ink in this felt-tip pen been made?

d) i) Is the yellow used in the orange pen the same as the one used in the green pen?

   Give a reason for your answer.

   ii) How many yellow dyes have been used in the eight pens?

e) Which of the dyes is likely to be the least soluble in water? Give a reason for your answer.

2 You are supplied with

- a paper towel
- a piece of jotter paper
- a piece of newspaper
- four 250 cm³ beakers
- a piece of shiny notebook paper
- a supply of water.

Describe, with the aid of a diagram, how you would find out which of the four types of paper would be the best for a chromatography experiment.

11 Metals and non-metals

» Metals

Worked example

Boron is an element with a very high melting point. It is a poor conductor of both electricity and heat, and it is also brittle.

Suggest whether boron is a metal or a non-metal. Explain your answer.

Boron is probably a non-metal. This is because it is a poor conductor of both electricity and heat – both properties of non-metals. The high melting point is of no use in deciding, because some non-metals (e.g. carbon) and most metals have high melting points.

> **Hint**
>
> When asked to make a prediction and offer an explanation, give the prediction first and then give the explanation in a separate sentence.

Know

1 State what is meant by each of the following terms:

   a) malleable   b) ductile   c) brittle.

2 The following is a list of elements: calcium, carbon, copper, mercury, sulfur, iron, aluminium, bromine. State which elements are metals and which are non-metals.

3 Which of the following lists contains three magnetic elements?

   A gold, iron and zinc

   B cobalt, iron and nickel

   C aluminium, cobalt and silver

   D iron, nickel and tin

4 a) Give the names of the two elements that are liquids at room temperature (20 °C).

   b) Which of these two elements is a metal and which is a non-metal?

Apply

1 The uses of aluminium are related to its properties. Copy and complete the table with a different property for each use. The first one has been done for you.

Use	Property
Drinks cans	Easily moulded
Aeroplanes	
Window frames	
Pans for cooking food	
Overhead power cables	

Extend

1 Aluminium and iron have similar properties. Both metals

- are malleable and ductile

- are good conductors of electricity and heat

- have a high melting point.

   a) **i)** Choose two properties from the list that make iron a suitable metal for cooking pans.

   ii) Choose two properties from the list that make aluminium a suitable metal for power cables.

   b) Steel is an alloy containing iron.

   The differences between steel and aluminium are:

   - steel can rust, but aluminium resists corrosion

   - steel has a higher density than aluminium

   - steel is much stronger than aluminium

   Use the information from the list above to suggest why

   i) steel is the better metal for making bridges

   ii) aluminium is the better metal for making aircraft bodies.

» Metals, acids and the reactivity series

Worked example

When a mixture of black copper oxide and black carbon is heated in a test tube, a pink-brown solid is formed.

a) Identify the pink-brown solid.
b) Write a word equation for the reaction.
c) Explain which element, copper or carbon, is the more reactive.
d) When this reaction takes place on a crucible lid, very little of the pink-brown solid can be seen. Explain why.

a) Copper
b) copper oxide + carbon → copper + carbon dioxide
c) Carbon is the more reactive element because it displaces copper from copper oxide.
d) Oxygen from the air reacts with the copper to reform copper oxide.

Know

1 What is the name of the reaction in which a more reactive metal takes the place of a less reactive metal in a compound?

   A displacement   **C** oxidation

   B neutralisation   **D** reduction

2 Which of the following word equations represents the reaction between zinc and sulfuric acid?

   A zinc + sulfuric acid → zinc sulfate + hydrogen

   B zinc + sulfuric acid → zinc sulfide + hydrogen

   C zinc + sulfuric acid → zinc sulfate + water

   D zinc + sulfuric acid → zinc sulfide + water

Apply

1 A student observes the reaction of dilute sulfuric acid with four metals, W, X, Y and Z. The table shows her observations.

Metal	Observations
W	Few bubbles produced slowly
X	Many bubbles produced quickly
Y	Many bubbles produced very quickly
Z	Very few bubbles produced very slowly

a) Use the information in the table to place the four metals in order of reactivity. Place the most reactive metal first.

b) The gas given off when a metal reacts with sulfuric acid is hydrogen. Describe a test to show that the gas is hydrogen.

> **Hint**
>
> When describing a test you should also give the result of the test if positive.

2 The reactivity of some metals can be compared by their reactions with dilute hydrochloric acid.

Zinc, iron and magnesium are added to separate test tubes containing dilute hydrochloric acid.

The diagram shows bubbles of hydrogen gas forming when a piece of zinc is added to dilute hydrochloric acid.

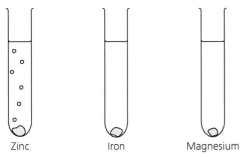

Zinc   Iron   Magnesium

a) Copy the diagram and complete it to show the bubbles forming in the other two test tubes.

b) Write a word equation for the reaction between zinc and hydrochloric acid.

Extend

1 The list gives the order of reactivity of some metals.

Most reactive ↑ Magnesium
Zinc
Iron
Copper
Least reactive Silver

A student is given some solid nickel nitrate and several small pieces of magnesium, zinc, iron, copper and silver. Explain how she can find the position of nickel in the reactivity series given above.

2 Some magnesium powder is added to dilute hydrochloric acid in a test tube.

A colourless solution is formed and a gas is given off.

When more magnesium powder is added, the reaction continues for a while and then stops.

Some magnesium powder is left at the bottom of the test tube.

When a lighted spill is placed at the mouth of the test tube, the gas burns with a squeaky pop.

a) Identify the gas given off.

b) Suggest why the reaction stops.

c) Give the name of the colourless solution.

3 Uranium is a metal that is in between magnesium and zinc in the reactivity series.

Equal sized pieces of magnesium, zinc and uranium are placed in separate solutions of dilute hydrochloric acid.

The observations for magnesium and zinc are shown in the table.

Metal	Observations
Magnesium	Bubbles of gas are produced rapidly. Solid disappears very quickly.
Zinc	Bubbles of gas are produced slowly. Solid disappears slowly.

a) State two variables the students should keep the same to make sure that the experiment is a fair test.

b) Suggest the observations that would be made for uranium.

c) When a piece of burning magnesium ribbon is lowered into a gas jar containing carbon dioxide, the magnesium continues to burn. A white solid and a black solid are formed.

The equation for the reaction is:

magnesium + carbon dioxide → magnesium oxide + carbon

i) Identify the white and black solids.

ii) Explain whether magnesium or carbon is the more reactive element.

iii) Explain which element is being oxidised in this reaction.

» How do metals and non-metals react with oxygen?

Worked example

When a piece of burning calcium is lowered into a gas jar containing carbon dioxide gas, the calcium continues to burn. A white solid and a black solid are formed. Explain these observations.

The calcium reacts with the carbon dioxide forming calcium oxide, a white solid and carbon (a black solid). The reaction takes place because calcium is more reactive than carbon.

> **Hint** !
>
> **When asked to explain observations of a reaction, it is important to state what has caused the observations, and also why the reaction has taken place.**

Know

1 What is the name of the reaction in which an element combines with oxygen to form an oxide?

  A displacement

  B neutralisation

  C oxidation

  D reduction

2 State what is seen when a piece of magnesium ribbon is burned in oxygen.

Apply

1 Nickel does not react with zinc oxide when heated together. Nickel reacts with lead oxide when heated together to form nickel oxide and lead.

  a) Place the elements lead, nickel and zinc in order of their reactivity. Place the most reactive first.

  b) Explain what reaction, if any, will take place when:

    i) a mixture of lead and nickel oxide is heated

    ii) a mixture of zinc and nickel oxide is heated.

2 Sapphires are made of an aluminium compound with the formula Al_2O_3. The chemical symbol for aluminium is Al.

  a) i) Give the name of the other element combined with aluminium in this compound.

    ii) Suggest the name of the compound with the formula Al_2O_3.

Sapphires are often mounted in gold to make rings. Gold is an element found in rocks. Gold is hardly ever found combined with other elements. Part of the reactivity series is shown below:

Most reactive	Aluminium
↑	Zinc
	Iron
	Lead
Least reactive	Copper

  b) Suggest where gold should be placed in this reactivity series.

Extend

1 A student notices that his bicycle rusts more quickly when he leaves it out in the open, especially during the winter when there is salt on the road. He decides that rust is caused by air, salt and water reacting with iron.

To test his theory, he puts an iron nail in a test tube containing air and some salt water. The iron nail rusts.

Next, he repeats the experiment without the salt. Again, the iron nail rusts.

a) What two deductions can be made about the role of salt in the rusting process?

The student then boils some water in order to remove the dissolved air. He puts an iron nail into this water and corks the test tube. This time the iron nail does not rust.

b) What does this result suggest about the role of air in the rusting process?

He then puts an iron nail into a test tube that contains a substance that absorbs the water vapour in the air. Once again, the iron nail does not rust.

c) What does this result suggest about the role of water in the rusting process?

In the final experiment, the student boils some water, adds a nail and then passes nitrogen gas into the water and corks the tube. The nail does not rust.

d) i) Suggest what substances must be present in order for iron to rust.

ii) Describe an experiment you could do to test your idea.

2 A council waste incinerator will produce fumes of hydrochloric acid if the plastic PVC is burned in it. The council claim that they have managed to stop these fumes from getting into the surrounding air. They support this claim by stating that a copper roof of a nearby building has not turned green, which is the colour of copper chloride.

a) How would you explain to the council that this does not prove anything?

b) Would it make a difference to the council's argument if the roof were made of iron?

» Displacement reactions

Worked example

Iron displaces copper from a solution of copper sulfate. Magnesium displaces iron from a solution of iron sulfate. Write a word equation for each of these two reactions and put the metals in order of reactivity (i.e. most reactive first).

iron + copper sulfate → iron sulfate + copper
magnesium + iron sulfate → magnesium sulfate + iron
magnesium > iron > copper

Know

1 What is the name of the reaction in which a metal takes the place of a less reactive metal in a compound?

A displacement

C oxidation

B neutralisation

D reduction

Apply

1 Use information from the table below to answer the questions that follow.

Increasing reactivity	Metal	Colour of solid metal	Colour of a solution of the metal sulfate
	Magnesium	Grey	Colourless
	Zinc	Grey	Colourless
	Iron	Dark grey	Green
	Copper	Pink-brown	Blue

a) Explain why no reaction occurs when zinc is added to magnesium sulfate solution.

b) When powdered iron is added to copper sulfate solution, a reaction takes place.

   i) Write a word equation for this reaction.

   ii) Describe the colour changes that occur during this reaction.

   Colour of solid changes from _____ to _____ .

   Colour of solution changes from _____ to _____ .

c) When copper is added to dilute sulfuric acid, no reaction occurs. When iron is added to dilute sulfuric acid, iron sulfate and hydrogen are formed. What does this suggest about the reactivity of hydrogen compared with the reactivity of copper and the reactivity of iron?

Extend

1 Some students are investigating displacement reactions involving three different metals – copper, zinc and metal X – and solutions of their salts. The equation below represents one of these reactions.

   zinc + copper sulfate solution → zinc sulfate solution + copper

This reaction occurs because zinc is more reactive than copper.

When a displacement reaction occurs, the temperature of the solution rises. The bigger the difference in reactivity between the two metals, the bigger the temperature rise.

The students use this method.

   (1) Pour some metal salt solution into a beaker, place a thermometer into the solution and measure the temperature.

   (2) Add some of the metal and stir the mixture.

   (3) Record the highest temperature reached.

a) State two variables the students should keep the same to make sure that the experiment is a fair test.

b) The table shows the results obtained.

Metal and metal salt used	Temperature rise in °C
Zinc + copper sulfate	12.7
X + copper sulfate	8.5
X + zinc sulfate	0.0
Copper + zinc sulfate	0.0
Zinc + X sulfate	2.9
Copper + X sulfate	0.0

i) Use these results to place the metals in order of their reactivity. Place the most reactive metal first. Give reasons for your answer.

ii) Write a word equation for the reaction between zinc and X sulfate.

iii) Explain why there was no temperature rise when copper is added to zinc sulfate solution.

2 In an experiment on displacement reactions involving metals and solutions of metal salts, the results shown in the table are obtained.

	Copper	Zinc	Magnesium	Silver	Iron	Lead
Copper nitrate solution	Not done	✓	✓	✗	✓	✓
Zinc nitrate solution	A	Not done	✓	✗	B	✗
Magnesium nitrate solution	✗	✗	Not done	✗	C	✗
Silver nitrate solution	✓	D	E	Not done	✓	✓
Iron nitrate solution	F	✓	✓	G	Not done	H
Lead nitrate solution	✗	I	J	✗	✓	Not done

✓ means the metal above displaced the metal from the salt solution named on the left.

✗ means the metal above did not displace the metal from the salt solution named on the left.

a) Use the results from the table to arrange the six metals in order of reactivity. Place the most reactive first.

b) State what results you would expect (i.e. a ✓ or a ✗) for each of the spaces labelled A–J.

c) State why copper was not added to copper nitrate solution.

12 Acids and alkalis

» The pH scale

Worked example

A solution has a pH of 6. What does this tell you about the solution?

A solution with a pH of 6 is weakly acidic.

Know

Hint

There are five deductions you can make about a solution from its pH value. They are: strongly acidic, weakly acidic, neutral, weakly alkaline, strongly alkaline.

1 Sulfuric acid is a strong acid. Citric acid is a weak acid. Sodium chloride is a neutral salt. Suggest a pH value for a dilute solution of each.

2 Give the chemical formula for each of the following acids and alkalis:

   a) hydrochloric acid

   b) sodium hydroxide

   c) sulfuric acid

   d) potassium hydroxide

   e) nitric acid

   f) calcium hydroxide.

3 The pH of a solution can be measured using a pH meter. Describe one other method to measure the pH of a solution.

4 Which of the following statements about bases is true?

   A They are all alkalis.

   B They can neutralise acids.

   C They are all soluble in water.

   D They react with an acid to produce hydrogen.

Apply

1 Phenol red is an indicator. Its colour at different pH values is shown in the table.

pH 4	pH 7	pH 10
Yellow	Orange	Red

Some dilute hydrochloric acid is placed in a beaker.

A few drops of phenol red indicator are added to the acid.

Dilute sodium hydroxide is added gradually until it is in excess.

   a) What colour is phenol red indicator in dilute hydrochloric acid?

   b) What colour is phenol red indicator when excess sodium hydroxide is added?

   c) What type of reaction takes place when hydrochloric acid reacts with sodium hydroxide?

   d) Give the name of a pure liquid in which phenol red indicator is orange.

2 The diagram gives the colours of some indicators at different pH values.

Indicator	pH						
	1	3	5	7	9	11	13
Litmus	← Red →			Purple	← Blue →		
Phenolphthalein	← Colourless →				← Pink →		
Methyl orange	← Red →		← Yellow →				

a) Use the table to find the pH of a solution in which litmus is red and methyl orange is yellow.

b) Litmus is purple in sodium chloride solution.

   What colour is phenolphthalein in sodium chloride solution?

Extend

1 The diagram shows the colours of three indicators at different pH values.

	pH 0 1 2 3 4 5 6 7 8 9 10 11 12 13 14
Litmus	← Red → Purple ← Blue →
Methyl orange	← Red → Orange ← Yellow →
Phenol red	← Yellow → Orange ← Red →

The table below shows the colours of the solutions of three compounds, carbon dioxide, ammonia and sulfur dioxide, in the three indicators.

Solution	Colour in litmus	Colour in methyl orange	Colour in phenol red
Ammonia	Blue	Yellow	Red
Carbon dioxide	Purple	Yellow	Yellow
Sulfur dioxide	Red	Red	Yellow

Using the information in the diagram and table, what can you deduce about the pH values of each solution?

2 Some universal indicator is added to 50 cm³ of potassium hydroxide in a conical flask. A total volume of 50 cm³ of dilute nitric acid is added gradually to the potassium hydroxide solution.

The graph shows how the pH of the solution changes as the acid is added.

Use the graph to answer these questions.

a) i) What is the pH of the potassium hydroxide solution before any acid is added?

   ii) What is the pH of the solution after 30 cm³ of acid has been added?

   iii) What volume of acid is needed to completely neutralise the potassium hydroxide?

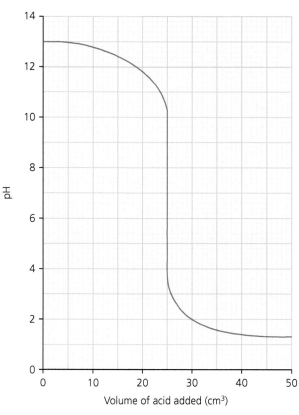

b) The table shows the colour of universal indicator at different pH values.

pH	0–2	3–4	5–6	7	8–9	10–12	13–14
Colour	Red	Orange	Yellow	Green	Blue	Indigo	Violet

What is the colour of the solution when:

i) 20 cm³ of nitric acid is added

ii) 35 cm³ of nitric acid is added.

c) Write a word equation for the reaction between nitric acid and potassium hydroxide.

» Acids reacting with alkalis

Worked example

An old memory aid for treating bee and wasp stings is '**B**ee – **B**icarb'; '**V**inegar – **V**asp'. 'Bicarb' is sodium hydrogencarbonate, a weak alkali. Vinegar is a weak acid.

What do these treatments suggest about the nature of bee and wasp stings? Justify your answer.

Bee stings may be acidic and wasp stings may be alkaline. This is assuming that the remedy involves a neutralisation reaction. The alkaline sodium hydrogencarbonate will neutralise the acid in the bee sting, while the acidic vinegar will neutralise the alkali in the wasp sting.

Know

1 What is the name of the type of reaction that takes place when an acid reacts with an alkali?

A displacement

B neutralisation

C oxidation

D reduction

2 State what is meant by the term indicator.

3 What is the colour of methyl orange in acidic solutions and in alkaline solutions?

Apply

1 The diagram shows part of the pH scale.

pH   0 ---------------------- 7 ---------------------- 14
Strongly acidic   Neutral   Strongly alkaline
   solution   solution

Some of these experiments involve a pH change.

A Carbon dioxide gas is dissolved in pure water.

B Excess sodium hydroxide solution is added to a weakly acidic solution.

C Sodium hydroxide solution is neutralised by adding citric acid.

D Sodium chloride (common salt) is dissolved in pure water.

E Hydrochloric acid is added to pure water.

The table shows the pH at the start and at the end of each of the five experiments. Copy and complete the table by inserting the appropriate letter in each box. The first one has been done for you.

pH at start	pH at end	Experiment
5	14	B
7	6	
7	7	
7	1	
14	7	

2 A vegetable has leaves that are normally green in colour, but which turn red when cooked with vinegar. Vinegar contains the weak acid, acetic acid.

a) State how you would try to extract and then separate the coloured substances in the green vegetable leaves.

b) How would you find out which of these coloured substances change colour when treated with vinegar?

c) What is the name given to the type of substance that changes colour in this way?

d) State how you could find out whether it is the cooking or the vinegar that turns the colour from green to red.

Extend

1 Dilute hydrochloric acid can be used to make salts. The salts produced are called chlorides.

a) Some sodium compounds react with hydrochloric acid to produce sodium chloride.

Which one of the following compounds will not react with hydrochloric acid to produce sodium chloride:

A sodium carbonate

B sodium hydroxide

C sodium oxide

D sodium sulfate?

b) Indigestion tablets neutralise excess hydrochloric acid in the stomach.

Two tablets are tested.

The table shows the cost of each tablet and the volume of hydrochloric acid it will neutralise.

Tablet	Cost of one tablet	Volume of hydrochloric acid neutralised by one tablet
A	2.6 p	36.0 cm³
B	1.3 p	12.0 cm³

Explain which tablet, A or B, is the best value for money.

2 A student wants to compare the acidity of four gases to find out which gas might cause acid rain.

She bubbles each gas separately through universal indicator solution.

Three of the gases cause the indicator to change its green colour.

The student adds an alkali drop by drop to each solution until the indicator changes back to green. She counts the number of drops required in each case.

The table shows her results.

Name of gas tested	Change in colour of indicator	Number of drops of alkali needed to change the indicator back to green
Carbon dioxide	Green To Yellow	30
Methane	No Change	0
Sulfur dioxide	Green To Orange	100
Nitrogen dioxide	Green To Red	150

a) State which variables the student should keep constant in order to make the experiment a fair test.

b) Which gas produces the most strongly acidic solution? Explain your choice.

c) Which gas appears to form a neutral solution? Explain your choice.

d) What effect does an alkali have on an acid?

3 A concentrated solution of a strong acid is split on the floor of the laboratory. Your teacher uses a weak alkali, sodium carbonate, to neutralise it.

a) How do you know when the acid has been completely neutralised?

b) Why is it not sensible to use a strong alkali to neutralise the acid?

c) Why is water not used in this case?

» Acids reacting with metal carbonates

Worked example

Write a word equation for the reaction between copper carbonate and nitric acid.

copper carbonate + nitric acid → copper nitrate + water + carbon dioxide

Know

1 What is the name of the gas produced when an acid reacts with a carbonate?

2 What products are formed when an acid reacts with a metal carbonate?

    A a salt only   **C** a salt and hydrogen

    B a salt, carbon dioxide and water   **D** a salt and water

3 Which of the following is the test for carbon dioxide?

    A It produces a pop when ignited.   **C** It turns damp red litmus blue.

    B It relights a glowing spill.   **D** It turns limewater milky.

Apply

1 The word equation for the reaction between zinc carbonate and dilute nitric acid is shown below:

zinc carbonate + nitric acid → zinc nitrate + carbon dioxide + water

The apparatus below is used to investigate this reaction.

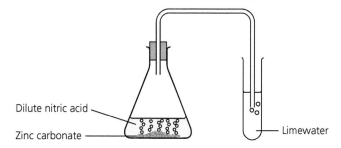

Dilute nitric acid

Zinc carbonate

Limewater

    a) Suggest a pH value for dilute nitric acid.

    b) What change do you observe in the limewater as the carbon dioxide bubbles through it?

    c) At the end of the reaction all of the nitric acid has been used up.

    d) Suggest a value for the pH of the remaining solution.

2 A student collected two different samples of rainwater and measured the pH of each sample. The table shows his results.

Sample	Where collected	pH
A	Rainwater collected as it fell from the sky	5.6
B	Rainwater collected after it had passed through limestone rocks	6.7

a) Which of the following statements about rainwater is true?

   A It is strongly acidic.   C It is strongly alkaline.

   B It is weakly acidic.   D It is weakly alkaline.

b) What kind of reaction occurs when rainwater passes through limestone rock?

c) When nitric acid is added to limestone rock, calcium nitrate and carbon dioxide are both produced. Name the chemical compound that must be present in the limestone rock.

Extend

1 The chemical formula for hydrochloric acid is HCl.

The chemical formula for sodium hydroxide is NaOH.

When hydrochloric acid and sodium hydroxide react together, two substances form. The name of one substance is sodium chloride.

a) Give the name of the other substance that forms when hydrochloric acid reacts with sodium hydroxide.

b) In an experiment, a student places two beakers on a balance.

The first beaker contains 20 cm³ of hydrochloric acid.

The second beaker contains 20 cm³ of sodium hydroxide.

The total mass of the two beakers and contents is 175.75 g.

The student then pours the hydrochloric acid into the sodium hydroxide. She then places both beakers on the balance. The reading on the balance is still 175.75 g.

Explain why the mass stays the same.

c) In a second experiment, the student again places two beakers on the balance.

The first beaker contains 20 cm³ of hydrochloric acid.

The second beaker contains 8.4 g of magnesium carbonate.

The total mass of the two beakers and contents is 135.64 g.

The student then pours the hydrochloric acid onto the magnesium carbonate and allows the reaction to complete. She then places both beakers on the balance. The reading on the balance is 131.24 g.

   i)   Write a word equation for the reaction that takes place.

   ii) How does the student know when the reaction is complete?

   iii) Explain why there is a loss in mass in this reaction.

13 Earth structure

» The structure of the Earth

Worked example

Feldspars make up approximately 60% of the Earth's crust. One type of feldspar has the formula $KAlSi_3O_8$. Is this feldspar a mineral or a rock? Explain your answer.

It is a mineral since it has a definite chemical formula.

Know

1 Describe each of the following parts of the Earth:

a) the core

b) the mantle

c) the crust.

2 What is the difference between most rocks and a mineral?

3 Why is coal not classed as a mineral?

Apply

1 A student is comparing the hardness of five rocks. These are her results.

- Rock A makes rock C crumble.
- Rock B makes rock A crumble.
- Rock C makes rock D crumble.
- Rock E makes rock B crumble.

Arrange the rocks in order of hardness. Put the hardest rock first.

Extend

1 a) What are the two main elements in the inner core of the Earth?

b) The inner core and the outer core of the Earth are made up of the same substances.

The temperature of the inner core is about 5400 °C.

The temperature of the outer core is about 3500 °C.

Explain why the inner core is liquid and the outer core is solid, despite the inner core being at a higher temperature.

» Types of rock

Worked example

The diagram shows an arrangement of rock layers in which three different types of rock, A, B and C, are present.

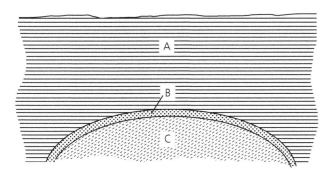

One rock is igneous (granite), another is metamorphic (quartzite) and the third is sedimentary (sandstone).

a) Use the information above to state which of the rocks is granite, which is quartzite and which is sandstone.

b) Describe how quartzite is formed.

a) A is sandstone; B is quartzite; C is granite

b) The action of heat and pressure on the sandstone

Know

1 a) State what is meant by the following terms:

   i)   sedimentary rock

   ii) igneous rock

   iii) metamorphic rock.

b) Give an example of each type of rock listed in part a.

2 The diagram shows the remains of a plant that lived millions of years ago.

Use words from the box to complete the sentences that follow.

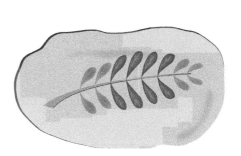

fossils	igneous	metamorphic
sedimentary	soil	

a) The remains of the plant are called _____ .

b) The remains of plants are found in _____ rocks.

c) Weathering can break down rocks to form _____ .

Apply

1 **a)** Explain how fossils were formed.

   b) Explain why fossils are not found in igneous or metamorphic rock.

   c) Explain why very few fossils are found of creatures with soft bodies.

2 The diagrams show three fossils commonly found in different layers of sedimentary rock.

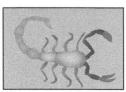  

| 280 million years ago | 345 million years ago | 430 million years ago |

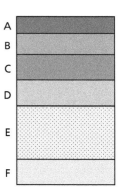

   a) Describe how sedimentary rocks are formed.

   b) Why are sedimentary rocks formed in layers?

The diagram on the right shows a sample of six layers of sedimentary rock.

   c) Fossil 2 above was found in layer E of the sample. In which layer might you expect to find fossil 3? Give a reason for your answer.

Extend

1 The table below shows the properties of some minerals.

Mineral	Formula	Relative density	Hardness
Calcite	$CaCO_3$	2.75	3.0
Diamond	C	3.5	10.0
Fluorspar	CaF_2	3.0	4.0
Gold	Au	12.0	3.0
Halite	NaCl	2.2	2.0
Sphalerite	ZnS	3.9	3.5

   a) Which of these minerals are elements?

   b) Name the elements present in calcite.

   c) Suggest which mineral is used to make sulfuric acid (H_2SO_4).

   d) Suggest which of the minerals is the best to use as part of tools for cutting.

2 Explain how some fossils can be used to date rocks.

» The rock cycle

Worked example

Weathering of limestone carvings leaves them pitted and worn away. The weathering is caused by the action of acid rain. Is this an example of chemical or physical weathering? Explain your answer.

Chemical weathering: the acid in the rain reacts with the limestone.

> **Hint**
>
> Chemical weathering involves a chemical reaction, whereas physical weathering does not.

Know

1 What is the difference between weathering and erosion?

2 What type of weathering is caused by the action of running water on rocks in streams or rivers?

Apply

1 a) State two ways in which rocks are broken down by weathering.

   b) After rocks have been broken down, the particles are sometimes carried away by natural processes.

      i) What name is given to the process in which the particles are carried away?

      ii) State one way in which the particles may be carried away.

   c) Which of the following rocks has been formed by the action of heat and pressure?

      A chalk

      B granite

      C limestone

      D marble

2 The diagram shows a rock being weathered until it eventually becomes rounded.

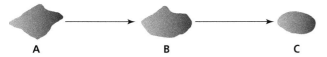

A   B   C

   a) Suggest two different ways in which the rock has become rounded.

   b) Suggest a place where rock C is most likely to be found. Give a reason for your answer.

3 The diagram below is a sketch of a quarry face.

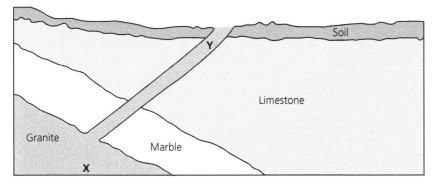

   a) The limestone was originally formed in layers. State what type of rock limestone is.

   b) Granite is an igneous rock formed from magma.

      i) Explain how the marble has formed above the granite.

      ii) Explain why the rock at Y contains smaller crystals than the rock at X.

4 The diagram shows different rock types found on an island as it was many years ago and as it is today.

Island many years ago

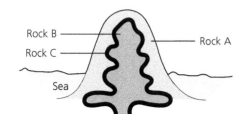

Rock B
Rock C
Rock A
Sea

Island today

Rock B
Rock C
Rock A

a) The diagram shows that rock A used to cover the top of the island. Copy and complete the paragraph that explains why the layer of rock A has changed. Choose words from this list:

erosion	transported	weathered	water

Wind and _____ have _____ rock A, breaking it into small pieces. The small pieces have then been _____ away by water and wind. This is called _____ .

b) Rock B is igneous rock.

   i) How is igneous rock formed?

   ii) Rock B was made of large crystals. What does this tell you about the way the rock was formed?

c) Rock C was formed from rock B by heating and pressure. What name is given to rock that is changed by heating and pressure?

Extend

1 The diagram represents part of the rock cycle.

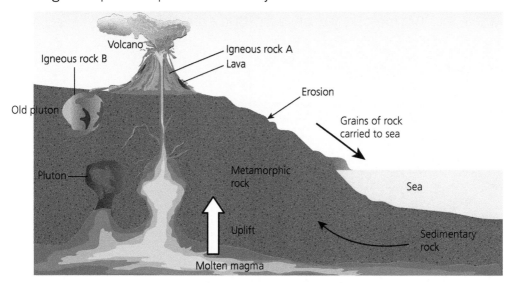

Volcano
Igneous rock B
Igneous rock A
Lava
Erosion
Old pluton
Grains of rock carried to sea
Pluton
Metamorphic rock
Sea
Uplift
Sedimentary rock
Molten magma

a) Explain how the grains of rock carried to the sea are changed into sedimentary rock.

b) Why do sedimentary rocks often contain fossils?

c) Explain why igneous rock A contains small crystals.

d) The diagram contains two plutons. A pluton is rock formed when molten magma is trapped before it reaches the surface.

   Explain why igneous rock B contains large crystals.

e) What two conditions are required to convert sedimentary rock into metamorphic rock?

2 The diagram below represents the rock cycle.

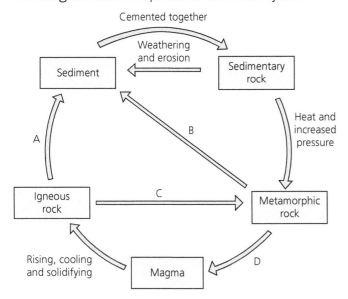

Describe the processes that occur at A, B, C and D.

3 Two cottages are built of stone.

One cottage is built using granite. The other is built using sandstone.

a) i) To which rock group does granite belong?

   ii) To which rock group does sandstone belong?

   iii) Explain which cottage is likely to be least affected by the weather.

b) Slate is often used to make roofing tiles.

   i) To which rock group does slate belong?

   ii) Explain why slate is a good material to use for making roof tiles.

14 Universe

» The Earth in space

Worked example

Why, on Earth, does the Sun rise in the east and set in the west?

It is because the Earth spins towards the east.

Know

1 State, in terms of the Earth's movement, what is meant by each of the following terms.

   a) an orbit   **b)** a year   **c)** a day

2 How long does it take for light to reach the Earth from:

   a) the Sun

   b) the Earth's nearest star?

   c) Give the name of the galaxy that contains our Solar System.

Apply

1 **a)** Spacecraft containing astronauts have reached the Moon but have so far not reached Mars. One reason for this is that:

   A the Moon is closer to the Earth than Mars

   B the Moon is closer to the Sun than Mars

   C Mars is closer to the Earth than the Moon

   D Mars is closer to the Sun than the Moon.

   b) Which of these objects in space is the smallest?

   A Earth   **C** the Moon

   B Jupiter   **D** the Sun

Extend

1 The Earth is one of eight planets in our Solar System.

   a) Give the names of two other planets in our Solar System.

   b) The diagram below represents the Sun, Earth and Moon.

Sun

Moon

P

Earth

A person at point P on the Earth can see the Moon.

i) Copy the diagram and draw lines to show how light from the Sun allows the person at point P to see the Moon.

ii) A lunar eclipse occurs when the Moon moves into the Earth's shadow.

Write the letter L on your diagram to show a possible position of the Moon during a lunar eclipse.

c) This diagram shows some of the ways the Moon can appear against the background of a dark sky.

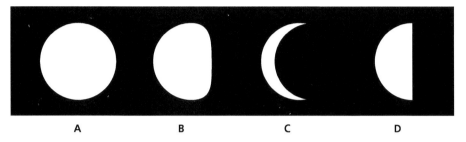

A	B	C	D

Which letter represents the view of the Moon that the person at P is likely to see?

2 At one time astronomers thought that the Earth was the centre of the Universe. This was called the geocentric model.

Astronomers now believe that the geocentric model is not correct.

The evidence for the geocentric model came from observations of the sky made using the naked eye.

The evidence against the geocentric model come from observations made with telescopes.

Describe the evidence both for and against the geocentric model.

» The Solar System

Worked example

Pluto spins on its axis in the opposite direction from the Earth. What effect would this have on an observation made on Pluto of the Sun's movement?

The Sun will rise in the west and set in the east.

Know

1 The diagram shows four moons that orbit Jupiter.

a) Jupiter is:

A a comet

B a galaxy

C a planet

D a universe.

b) Galileo used a new invention to observe these moons. Give the name of the invention he used.

Apply

1 The diagram shows the orbits of Ganymede and Europa around Jupiter.

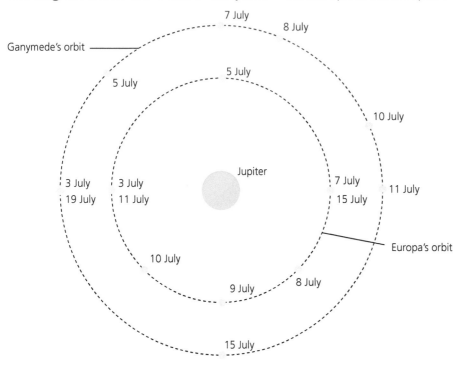

The positions of the two moons on some dates are marked.

a) i) On which one of the marked dates are the two moons closest together?

ii) How long does it take for Ganymede to complete one orbit of Jupiter?

b) The radius of Ganymede's orbit is 1 070 000 km.

The radius of Europa's orbit is 671 000 km.

Calculate the distance between Ganymede and Europa on 11 July.

c) Galileo used a telescope to observe Jupiter. His observations provided evidence to support the idea that the Earth is not the centre of the Universe.

Explain how Galileo's observations supported this idea.

2 The diagram shows the orbits of Earth, Neptune and Pluto. At the two points, marked X and Y, the orbits of Neptune and Pluto cross over.

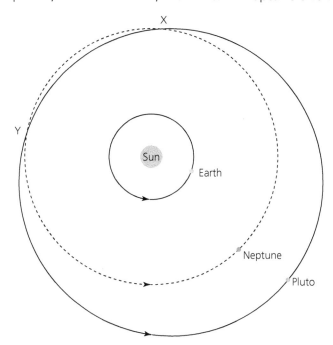

a) Give the name of the type of force that keeps planets in orbit around the Sun.

b) Give two reasons why it takes Pluto longer than Neptune to orbit the Sun.

c) i) Explain why Neptune and Pluto can be seen even though they do not give out their own light.

ii) Between points X and Y, Pluto is nearer than Neptune to the Earth.

Explain why Pluto does not appear to be as bright as Neptune, even when Pluto is closer than Neptune to the Earth.

Extend

1 The table gives information about the planets in our Solar System.

Planet	Diameter (km)	Atmosphere
Earth	12 800	Mainly nitrogen and oxygen
Jupiter	142 600	Mainly hydrogen and helium
Mars	6800	Mainly carbon dioxide
Mercury	4800	None
Neptune	50 000	Mainly hydrogen and helium
Saturn	120 000	Mainly hydrogen and helium
Uranus	49 000	Mainly hydrogen and helium
Venus	12 200	Mainly carbon dioxide

a) Pluto has a diameter of 2300 km.

How does this information support the view that Pluto is not a planet?

b) A moon called Charon orbits Pluto.

How does this information support the view that Pluto is a planet?

c) Atmosphere is not used to classify objects as planets. Suggest why.

d) Suggest why scientists found it difficult to decide how Pluto should be classified.

2 The diagram shows five phases of the Moon. They are in the wrong order in which they appear in the sky.

A   B   C   D   E

 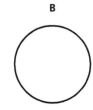   

Use the letters A, B, C, D and E to show the correct order.

3 The table below shows some information about four planets in our Solar System.

Planet	Approximate time taken to orbit the Earth (Earth years)	Distance from the Sun (millions of km)
Mercury	0.25	60
Venus	0.5	108
Earth	1.0	150
Mars	2.0	228

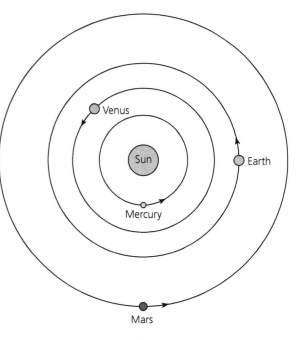

The diagram shows the orbits of the four planets and their position at a particular time.

a) Copy the diagram and mark on it the position of each planet six months later.

b) Use the information in the table to calculate the largest and smallest possible distance between the Earth and Mars.

c) The speed of light is 3 000 000 km/s.

Calculate how long it takes for light from the Sun to reach Mercury.

» Phases of the Moon

Worked example

The image shows a solar eclipse. A solar eclipse occurs when the Moon passes between the Sun and the Earth.

a) The Moon is much smaller than the Sun. Which one of the following explains why the Moon appears to be about the same size as the Sun during a solar eclipse?

   A The Moon is closer to the Earth than the Sun is.

   B The Sun is closer to the Earth than the Moon is.

   C The Moon goes around the Earth much faster than the Sun does.

   D The Sun goes around the Earth much faster than the Moon does.

b) The graph below shows how the light levels changed during the eclipse.

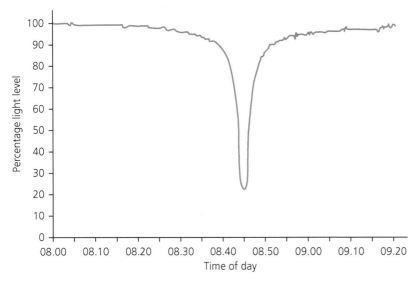

  i) At what time of day did the Moon block out most of the Sun's light?

  ii) Suggest what happens to the temperature of the air during the eclipse. Give a reason for your answer.

a) A (The Moon is closer to the Earth than the Sun is)

b) i) 08.45

   ii) The temperature drops because the Moon blocks out the heat from the Sun.

Know

1 The Moon does not give off light, so why can it be seen from the Earth?

2 What is the name given to the plane of the Earth's orbit around the Sun?

Apply

1 The diagram shows the eight-phase lunar cycle.

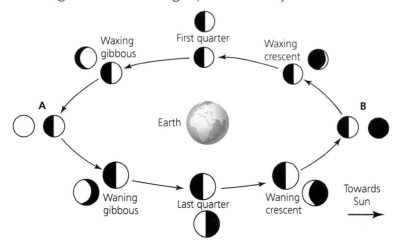

What name is given to the moon at point A and at point B?

Extend

1 The diagram shows the Moon's orbit around the Sun and rays of light coming from the Sun.

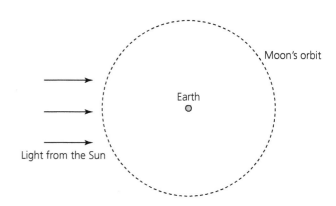

a) On a copy of the diagram, use the letter A to indicate the position of the Moon during a solar eclipse.

b) Use the letter B to indicate the time of a full Moon.

c) Draw a sketch of the appearance of the Moon during a partial solar eclipse.

d) Describe the appearance of the Moon during a lunar eclipse.

» Beyond the Solar System

Worked example

The distance from Earth to Alpha Centauri is 41 340 000 000 000 km. How long does it take for light from Alpha Centauri to reach Earth?

[The speed of light is 300 000 km/s]

time = distance/speed

= 41 340 000 000 000 km/300 000 km/s

= 137 800 000 s

(This is equivalent to 4.37 years – there are 31 536 000 seconds in one year. This means that as we look at Alpha Centauri, we are seeing it as it was 4.37 years ago, not as it is today.)

> **Hint**
>
> To calculate time you need to rearrange the equation 'distance = speed × time'.

Know

1 State what is meant by each of the following terms:

   a) a galaxy   b) a light year   c) a star.

2 a) A galaxy is a collection of:

   A asteroids   C planets

   B moons   D stars.

   b) Andromeda is just one of the many galaxies that form the:

   A constellations   C stars

   B planets   D Universe.

Apply

1 Scientists use telescopes to search for new planets orbiting distant stars. The Kepler space telescope orbits above the Earth's atmosphere and records the brightness of light from distant stars. When a planet passes in front of a distant star, there is a tiny dip in the brightness of the light from the star.

   a) Suggest why telescopes that search for planets orbiting distant stars are not on the Earth's surface.

   b) The Kepler telescope was pointed towards one star. It detected the same dip in the brightness of the star every 150 days. Suggest what information this gives about the planet that orbits this star.

Extend

1 Other than the Sun, the star closest to Earth is Proxima Centauri.

   Light from this star takes 2 200 000 minutes to reach the Earth.

   Light from the Sun takes 8.3 minutes to reach the Earth.

   The speed of light is 18 000 000 km/min.

   a) Calculate the distance of Proxima Centauri from the Earth.

   b) Calculate the distance of the Sun from the Earth.

   c) How many times further away from the Earth is Proxima Centauri than the Sun?

   d) A light year is the distance that light travels in one year.

   Suggest why astronomers usually give the distance of the stars from the Earth as a number of light years instead of a number of kilometres.

15 Movement

» Skeletons

Worked example

Explain why in humans, the pelvis of a woman tends to be wider than the pelvis of a man.

Because a wide pelvis makes it easier for a baby to be born as the baby has to be born through the pelvis. Men do not get pregnant and give birth, so they do not benefit from having a wider pelvis.

> **Hint**
> When answering an 'explain' question, the word 'because' is a really useful explaining word.

Know

1 Give the four functions of a skeleton.

2 What is the name of an area where two bones of a skeleton meet?

3 Skeletons and muscles allow animals to move. What is the name of the structure that joins muscles to bones?

4 Give the name of the part of some bones where blood cells are formed.

Apply

1 When babies are born the bones in their skulls are not fused together like they are in adults. This makes a baby's head soft and delicate. Give a reason why the bones in babies' skulls are not fused together.

2 The bones of birds are often hollow.

   a) Suggest a reason why this would be good for a bird.

   b) Suggest a reason why this might be bad for a bird.

> **Hint**
> Gravity is a force that pulls objects down toward the centre of a planet. The force of gravity acting on an object is called the object's weight. The greater the force of gravity, the greater the weight of the object.

Extend

1 Astronauts that spend a lot of time in the very low gravity (or microgravity) environment of the International Space Station (ISS) suffer from the loss of bone density and muscle atrophy (the wasting away of muscle tissue). Astronauts can lose up to 0.25 g per day of calcium from their bones, which is a significant amount. Unless astronauts exercise while in space, they can have serious health problems on their return to Earth.

   a) Calculate the mass of calcium that could be lost from an astronaut during a six month stay on the ISS.

   b) Suggest a reason why muscles and bones waste away in a microgravity environment like the ISS.

> **Hint**
> Why does your muscular–skeletal system have to work harder on Earth than it would in a microgravity environment?

c) Predict the health problems that an astronaut might have on returning to Earth having not exercised in space.

d) As well as exercise, astronauts take vitamins and minerals to help maintain their skeletons. Use the internet to find out which vitamin is essential for maintaining healthy bones, and describe how it keeps bones healthy.

» Joints

Worked example

Cartilage is a smooth tissue found at the end of bones. Explain the importance of cartilage. Try to give as much detail as you can.

Cartilage is important because it protects the ends of the bones in a joint. Without cartilage the bones would rub together and wear away. This would cause a lot of pain and cause arthritis. Cartilage is smooth so that it can reduce friction in the joint and allow the joint to move smoothly.

> **Hint**
>
> This question encourages you to go into detail. If you cannot remember what cartilage does, then look back through the pages of your Student Book to remind yourself.

Know

1 Copy and complete the following paragraph using your knowledge of joints. Use the words in the box below.

cartilage	contract	friction	
joint	ligaments	relax	tendons

The adult human skeleton is made up of 206 bones, and where any two bones meet is called a _____ . Pairs of muscles around a joint allow it to move because muscles are able to _____ (shorten) and _____ (lengthen). A joint has several different structures within it. For instance, _____ join muscle to bone, _____ connect one bone to another across a joint, and _____ is a smooth tissue found at the end of bones, which reduces _____ between them.

> **Hint**
>
> 'Justify' means to give an explanation for something. So in Extend Question 1 you need to explain why you chose your joint type.

Apply

1 When you curl your index finger, two joints in your finger move. Based on how your finger moves, what type of joint (pivot, hinge, or ball and socket) do you think is in your fingers? Justify your answer.

2 Explain, using your knowledge of how your shoulder moves, why the joint in a shoulder is a ball and socket joint rather than a hinge joint.

3 There are 22 bones, and many joints, in an adult human skull. Almost none of these joints can move; they are fixed joints.

   a) Which part of your skull can actually move?

   b) Suggest what would happen to the shape of the skull if it contained moving joints.

   c) Based on your answer to part b, suggest a reason why the skull contains so many fixed joints.

> **Hint**
>
> Question 2 gives a clear instruction about how you need to answer it. Your answer must include some information about how a shoulder can move.

Extend

1 Knee injuries in sports like football and rugby are very common, and often involve damage to the ligaments within the knee. Ligaments can be over-stretched, partially torn or completely torn.

   a) What is the role of ligaments in a joint?

   b) Predict the symptoms an athlete might experience if he or she

      i) over-stretches a ligament in the knee

      ii) partially tears a ligament in the knee

      iii) completely tears a ligament in the knee.

2 Evaluate the effectiveness of cartilage in synovial joints.

Hint

Focus your answer on the amount of pain, swelling and movement of the knee.

Hint

To 'evaluate' means to weigh up the advantages and disadvantages of something.

» Antagonistic muscle pairs

Worked example

Describe in detail what happens to your quadriceps and hamstring (an antagonistic muscle pair) when you straighten your leg in front of you while sitting down.

When you are sitting down, your leg is bent. The quadriceps muscles are relaxed and long but the hamstring is contracted and short. To straighten your leg the quadriceps must contract and shorten while the hamstring relaxes and lengthens.

Hint

This answer contains key terms used in your Student Book, like 'contract' and 'relaxes'. Try to use words like these in your answers too (when appropriate).

Know

1 What is an antagonistic muscle pair?

2 Give two examples of antagonistic muscle pairs in the human body.

3 What two actions can muscles do that result in movement?

Apply

1 The muscles at the back of your lower leg are called calf muscles and the muscles at the front are called shin muscles. The calf and shin muscles are an antagonistic muscle pair that control the movement of your foot.

   a) When you lift your foot so that your toes are pointing upward, which muscle is contracting and which is relaxing? Explain your answer.

   b) Describe what happens to the calf and shin muscles when you go on tiptoes.

2 Muscles are not just for moving bones at joints. The heart is a muscular organ that has a very important role. Why must the heart be made from muscle?

Hint

Try to explain your answer using ideas about lengthening and shortening of the two muscles.

Hint

If you are not sure, try to find out what the heart does on the internet and see if this helps.

Extend

1 The Achilles tendon connects the calf muscle to the heel bone. It is the largest and most exposed tendon in the human body. Injuries to this tendon are common, including the complete tearing of the tendon.

a) What is the role of tendons in a joint?

b) Predict the effect of a complete tear of the Achilles tendon on the movement of the foot.

2 Crocodiles can generate the largest downward biting force of any animal, up to 16 460 newtons of force (compared with only 890 newtons in a human bite). But an elastic band can be used to keep a crocodile's mouth closed.

a) What does this tell you about the muscle pairs that open and close crocodile mouths?

b) Why do you think crocodile jaws have evolved in this way?

Hint

Remember that the calf muscle is one half of an antagonistic pair of muscles (see Question 1).

» Technology for improving human movement

Extend

1 American military companies are developing automated exoskeleton suits that can be used to carry large loads in extreme conditions like high-altitude or high-heat environments. The suits are worn around the user and strap onto their back (like a rucksack) and legs. The suits are controlled by powerful computers that sense the user's movements and adapt to them. They are powered by batteries.

a) What is an exoskeleton?

b) The military exoskeleton suits are made from titanium. Suggest the properties of titanium that make it a good choice for these suits.

c) Suggest some limitations to the use of exoskeleton suits.

d) A lot of money has been spent researching and developing these exoskeleton suits, and a lot of people believe that it is not ethically correct to spend this money in this way.

i) Other than for military uses, how else could this sort of technology be used?

ii) Science cannot answer ethical questions like 'Should exoskeleton suits be used for military purposes?'. Explain why.

2 Like endoskeletons, exoskeletons provide protection and support for the organism. But unlike endoskeletons, exoskeletons do not grow and have to be shed and re-grown, to allow the organism to grow larger itself. Having an exoskeleton also limits how large the organism can grow because of the weight of the exoskeleton itself.

a) Why is having to shed and re-grow an exoskeleton a disadvantage for an organism?

b) Coconut crabs are believed to be the largest organisms with an exoskeleton, while African elephants are the largest land organisms with endoskeletons. Use the internet to find out how many times heavier an African elephant can grow compared to a coconut crab.

16 Cells

» Comparing cells

Worked example

The cell membrane is a very important cell structure. Describe the function of the cell membrane, using key words where appropriate.

The cell membrane surrounds the cell and keeps the contents of the cell in one place. It also allows substances to diffuse into and out of the cell, without which the cell would die.

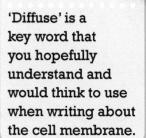

Hint

'Diffuse' is a key word that you hopefully understand and would think to use when writing about the cell membrane.

Know

1 Cells contain parts that can carry out the seven life process. What are the seven life processes?

2 Give the name of the part of a cell that contains genetic material, or DNA.

3 Animal cells and plant cells both contain cytoplasm. What is cytoplasm and why is it important?

Apply

1 Suggest two substances that might diffuse either into or out of a cell through the cell membrane. Explain your choices.

2 In plants, chloroplasts are mostly concentrated in the cells of leaves. Suggest a reason for this.

Extend

1 Explain why animal cells do not need to contain chloroplasts.

2 The type of microscope you may have used to see some cells is called an optical microscope. They use two magnifying lenses to view the object on the stage. The objective lens is close to the stage, while the eyepiece lens is further away. A microscope usually has three objective lenses with different magnifications. The image that is created is inverted, or upside down.

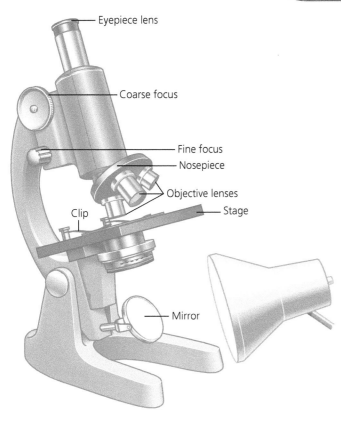

a) Draw the image you would see if you observed the word 'cell' using an optical microscope.

b) The image of a nucleus of an onion cell appears to be 2.4 mm across. It is viewed through an objective lens that magnifies by 40 times and an eyepiece lens that magnifies by 10 times. What is the actual size of the nucleus in millimetres?

» Specialised animal cells

Worked example

The female egg cell (or ovum) in humans is very large in comparison to the sperm cell (it can be seen without the aid of a microscope). Suggest reasons for this structural adaptation.

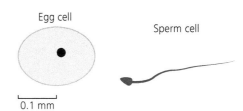

The egg cell is large as it contains lots of cytoplasm that stores energy sources for the growing embryo. A large egg cell is also easier for the smaller sperm cells to find and fertilise.

Know

1 You will have studied several specialised animal cells.

a) What cells have a branching structure to allow them to communicate with similar cells?

b) What cells have no nucleus to maximise how much oxygen they can transport?

c) What cells contain a lot of mitochondria to release lots of energy through respiration?

Apply

1 Some human sperm cells have deformities, including shorter tail sections or a smaller tail base.

   a) Suggest and explain how a shorter tail might affect the sperm's ability to fertilise an egg cell.

   b) Suggest and explain how a smaller tail base might affect the sperm's ability to fertilise an egg cell.

2 Ciliated cells in the human body are used to remove dust from the lungs and to waft the egg cell through the female reproductive system. Some unicellular organisms (organisms made up of only one cell) also have rows of cilia on the surface of their cell. Suggest a reason for this structural adaptation in unicellular organisms.

Extend

1 The total sperm count for a man is the total number of sperms produced during ejaculation. There is a relationship between the total sperm count of the man and the likelihood of the woman becoming pregnant, as shown in the table below.

Total sperm count (millions)	Pregnancy rate (%)	Total sperm count (millions)	Pregnancy rate (%)
0	0.0	15	15.0
2	4.0	20	16.5
4	8.0	25	18.0
6	10.0	30	18.3
8	12.0	35	18.8
10	13.0	40	19.0

   a) Plot a graph of sperm count (in millions) on the x-axis and pregnancy rate (in %) on the y-axis.

   b) Draw a smooth curve of best fit through your data points.

   c) Describe the pattern that your graph shows.

> Hint
>
> Remember to label each axis with the variable you are plotting on it, including the units of measurement.

» Specialised plant cells

Worked example

Explain why palisade cells are found in leaves, but not in the roots of a plant.

Palisade cells are specialised for photosynthesis – they have lots of chloroplasts that are necessary for photosynthesis to take place. Leaves are exposed to sunlight, but roots aren't, and sunlight is also essential for photosynthesis.

Know

1 What type of cells have structural adaptations to help them to absorb water and nutrients?

2 What type of cells have structural adaptations to help them to transport sugar solution around a plant?

Apply

1 Xylem cells are used to transport water through the plant. They have structural adaptations to help them do this job.

a) What are structural adaptations?

b) Give two structural adaptations of xylem cells.

c) Explain how each of these two structural adaptations help xylem cells to do their job.

2 Some plant cells (called sclerenchyma cells) have the job of supporting the plant as it grows. These cells help to keep the plant upright. They have a second cell wall that is strengthened by lignin.

a) From the information above, give the structural adaptation of sclerenchyma cells.

b) Suggest where in a plant you might find sclerenchyma cells: leaves, roots, flowers or stems?

c) What other specialised plant cell has a similar structural adaptation to sclerenchyma cells?

3 Not only palisade cells contain chloroplasts. Chlorenchyma cells also contain chloroplasts.

a) What function can chlorenchyma cells carry out?

b) Other than in leaves, where else might you expect to find chlorenchyma cells? Explain your answer.

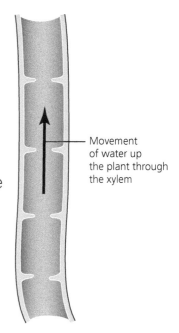

Movement of water up the plant through the xylem

Hint

Don't panic! You do not need to know or be able to pronounce the name of these cells.

Extend

1 The vacuole of plant cells is a very interesting structure that can have a variety of functions. Some functions of vacuoles include containing the waste products (produced in the cytoplasm) of the cell and removing them from the cell, and pushing all the other cell structures, including chloroplasts, outwards, toward the cell membrane.

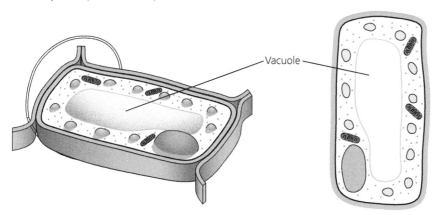

Vacuole

a) Given the information above, in which direction(s) will waste products move through the membrane of a vacuole: from the cytoplasm into the vacuole, from the vacuole into the cytoplasm, or both? Explain your answer.

b) What is the benefit of pushing chloroplasts outwards against the cell membrane?

Hint

Think about what chloroplasts do and what they need to do it.

» Organising cells

Worked example

An unknown specialised animal cell groups together to form tissue that is able to contract and relax by shortening and lengthening itself. Suggest what type of cell this animal cell might be and in what organ system it is most likely to be found.

It must be a muscle cell as muscle cells would form muscle tissue, and muscles are able to contract (shorten) and relax (lengthen). The organ system is most likely to be the muscular–skeletal system, which supports, moves and protects an organism.

Know

1 Define the following terms:

   a) unicellular organism

   b) multicellular organism.

2 Which human organ system is responsible for

   a) protecting the body against infections

   b) producing sperm and eggs, and supporting a growing fetus

   c) supporting, moving and protecting the body

   d) replacing oxygen and removing carbon dioxide from the blood?

Apply

1 Multicellular organisms have organ systems to carry out the seven life processes, while unicellular organisms, like bacteria, have to manage with just one cell.

   a) What structure of a bacterium helps it to move?

   b) What structure of a bacterium helps it to reproduce?

2 Copy and complete the following paragraph about the need for organ systems in multicellular organisms.

   Multicellular organisms need organ systems in order to carry out the seven _____ _____. In humans, _____ is achieved by the muscular–skeletal and nervous system; _____ is achieved by the reproductive system; _____ is carried out by the respiratory and circulatory system; and _____ and _____ are carried out by the digestive system.

Extend

1 Organ systems within an organism rarely work completely independently of one another; they are interdependent. If an organ in one organ system fails, different organ systems can also be affected.

The digestive system contains many organs, including the stomach, pancreas and intestines.

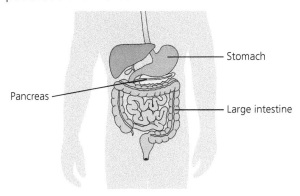

Pancreas

Stomach

Large intestine

a) What is the function of the digestive system?

b) What other organ system is needed for the digestive system to be able to operate? Explain your answer.

2 The small intestine forms part of the digestive system and is responsible for 90% of the absorption of nutrients into the body.

Describe and explain how the failure of the small intestine would affect the muscular–skeletal system.

» Healthcare and the human body

Extend

1 Penicillin was discovered by Alexander Fleming. It was, and is one of the first antibiotics to be used to fight bacterial infections. The core structure of penicillin has the chemical formula $C_9H_{11}N_2O_4S$. It works by weakening the cell wall of the bacteria, which results in the bacteria bursting open and dying.

a) What elements are present in a molecule of penicillin?

b) How many atoms in total are there in the core structure of penicillin?

c) Are bacteria living or non-living? Explain your answer.

2 Bacteria treated with penicillin burst open, because water is able to move into the cell through the weakened cell wall by osmosis. Use the internet to find out what osmosis is.

> **Hint**
>
> Use a periodic table if you don't recognise the chemical symbols in the formula for penicillin.

> **Hint**
>
> To help find websites that you can understand more easily, include "GCSE" in your search.

3 The nervous system is an organ system in the human body that controls movement and allows the body to respond to the world around it. Organs in this system include the brain, the spinal column and the nervous tissue. Messages are sent through nerve cells, or neurons, around our bodies.

Alcohol is a recreational drug that slows down the activity of the nervous system. Common symptoms of being drunk include slurred speech, an inability to carry out complex tasks like driving a car, and a loss of balance.

Based on these symptoms, what other organ system (other than the nervous system) is affected by alcohol?

4 Drinking a lot of alcohol can cause alcohol poisoning, which can result in death. One symptom of alcohol poisoning is a very slow breathing rate (fewer than eight breaths per minute).

a) What gas is absorbed into the bloodstream during breathing?

b) Which cells transport this gas around the body?

c) What does the body use this gas for?

d) What organ systems will be affected by a reduction in the amount of this gas in the body?

17 Interdependence

» Feeding relationships

Worked example

Describe the effect on a food chain of removing all the decomposers from an ecosystem, giving as much detail as you can.

Decomposers break down dead plant and animal matter, so that the nutrients they contain can be recycled by new plants. Without decomposers plants would have fewer of the nutrients they need to grow. This would reduce the size of the producer trophic level, and then reduce the size of all the other trophic levels until the food chain collapsed.

Know

1 Give the term for the following:

   a) a structure that shows how food chains in an ecosystem are linked

   b) an animal that eats other animals or plants

   c) a green plant or alga that makes its own food using sunlight.

Apply

1 Humans grow food, eat food and often compost our food waste so that the nutrients it contains can be used to grow other plants. Are humans producers, consumers or decomposers? Explain your answer.

2 How high up a food chain or web would you expect humans to be? Explain your answer.

3 What would happen to the number of trophic levels in a food chain if more energy were transferred between trophic levels? Explain your answer.

Extend

1 Use the internet to find out what the terms 'autotroph' and 'heterotroph' mean, and use this information to answer the following questions.

   a) Where would you find autotrophs in a food chain?

   b) Where would you find heterotrophs in a food chain?

   c) Are decomposers autotrophs or heterotrophs? Explain your answer.

2 Rabbits in Australia are an example of an invasive species: one that is introduced to an ecosystem and that tends to harm it. Rabbits are herbivores and breed quickly. They were introduced to Australia by European settlers in the 18th century. Suggest what happened to the food webs in the Australian ecosystems where rabbits were introduced.

» Pyramids of number and biomass

Worked example

What are the similarities and differences between a food chain and a pyramid of numbers?

Food chains and pyramids show which organisms consume others in an ecosystem, and they also show which way energy flows through the ecosystem. But a pyramid of number gives information about the numbers of living things in each population at each trophic level.

Know

1 Copy and complete the following paragraph using your knowledge of pyramids of number (use the words in the box).

chain	decreases	energy	first	fleas
imperfect	organisms	perfect	plant	

Pyramids of number show the number of _____ in each trophic level of a food _____. Usually, as we move up through the food chain the number of organisms _____ quickly because only 10% of the _____ of each trophic level is transferred to the next one and fewer organisms can survive. So we get a _____ triangle shape. Sometimes we get an _____ triangle shape; this happens when one very large _____, like a tree, forms the _____ trophic level. It can also happen when many parasitic organisms like _____ feed off the top predator in a food chain, causing a very wide top bar.

Apply

1 Two students are arguing about which type of pyramid is best: pyramids of number or pyramids of biomass.

a) What is biomass?

b) Give an advantage of pyramids of biomass over pyramids of number.

c) Give a disadvantage of pyramids of biomass.

2 Only about 10% of the energy of each trophic level is transferred to the next one. What happens to the other 90% of the energy in a trophic level?

Extend

1 In a typical food chain, there are 9000 kilocalories of energy per metre squared per year available from the producers in the first trophic level.

a) How much energy would be available to a tertiary consumer in this food chain if only 10% of the energy is transferred between each trophic level?

b) Show that there would be eight times as much energy available to the tertiary consumer if the amount of energy transferred between each level were doubled to 20%.

2 Venus flytraps are plants that also consume small insects like spiders and flies. They tend to grow in soils that have low levels of nutrients.

   a) Use the internet to find out why Venus fly-traps consume insects.

   b) How would you class Venus fly-traps: producers or consumers?

» Changes in ecosystems

Worked example

What is interdependence and why does it happen?

Interdependence is the fact that all organisms in an ecosystem depend on each other, even if they are not in the same food chain. It happens because food chains are linked together into food webs, so a change to one organism in one trophic level can affect many other organisms across the food web.

Know

1 Copy and complete the following paragraph using your knowledge of predator-prey cycles (use the words in the box).

fall	more	predators	prey
rise	smaller	transferred	trophic

_____ are animals that hunt and eat other animals and _____ are the animals that are hunted. An example of a predator–prey relationship in the UK is barn owls and voles. In any ecosystem there are usually _____ prey than predators. This links to what we know about pyramids of number and biomass; each new _____ level gets _____ as we go up the pyramid because only about 10% of the energy in each level is _____ to the next. If a change in the ecosystem reduces the number of prey, then the number of predators will _____ as they have less food. If the change causes the number of predators to fall, then the number of prey will _____ as fewer of them are being hunted and killed.

Apply

1 A simple marine predator–prey pair are seals and sardines. The sardines feed on plankton (small organisms that float in the ocean). Seals are hunted by polar bears.

   a) Describe and explain the effect of the loss of plankton on the seal population in this ecosystem.

   b) Describe and explain the effect of a fall in polar bear numbers on the sardine population.

   c) In another food chain in the same ecosystem, orcas (killer whales) hunt and eat sardines. Describe and explain the effect of a rise in polar bear numbers on the orca population.

Extend

1 Mountain lions in Canada play a vital role in maintaining their ecosystem. They range over a huge area and control the population sizes of their prey (deer, rabbits and other small mammals).

   a) Explain why the loss of one mountain lion can result in a large rise in deer and rabbit populations

   b) After the initial increase in deer and rabbit populations, a dramatic fall in the number of deer and rabbits (both herbivores) can occur. Explain why.

   c) Vultures are a scavenger species, and they are also controlled by the mountain lion. Explain how.

> **Hint**
>
> **Think about the effect on the trophic level below rabbits and deer.**

» Human impact on ecosystems

Extend

1 In 1962, Rachel Carson wrote a book called 'Silent Spring' that detailed the effect that DDT had on the environment. Many people believe that the publication of this book helped to cause the ban on the use of DDT in the USA in 1972. Explain how the publication of this book may have caused the banning of DDT.

2 Use the internet to find out how tobacco hornworms use bioaccumulation of a toxic substance to their advantage.

3 Strontium-90 is an isotope of strontium that is a product of atomic bombs and nuclear reactor disasters like Chernobyl. Strontium-90 is radioactive and can be bioaccumulated by organisms because it behaves a lot like calcium.

   a) Explain why strontium behaves like calcium.

   b) In which part of an organism would you expect to find strontium-90 as a result of bioaccumulation? Explain your answer.

   c) What diseases can radioactive elements cause?

   d) The approximate half-life for strontium-90 in organisms is 18 years. What does this mean and why is this a problem?

> **Hint**
>
> **Question 3a is really a Chemistry question!**

> **Hint**
>
> **You might need to use the internet to define 'half-life'.**

18 Plant reproduction

» Sexual reproduction

Worked example

Explain why plants benefit from reproducing sexually (i.e. with other plants in the same species).

By reproducing sexually, the new plants are likely to show greater variation, and this causes the gene pool of the plant species to widen. These plants are more likely to be able to survive a change to their ecosystem that might otherwise have made them extinct.

Know

1 Use the diagram below and your knowledge to give the terms for

 a) the joining of a nucleus from a male and female sex cell

 b) the structure from where pollen are released

 c) the structure that the ovary becomes after fertilisation, which contains seeds

 d) the structure where pollen grains land during pollination

 e) the structure that contains the embryo of a new plant.

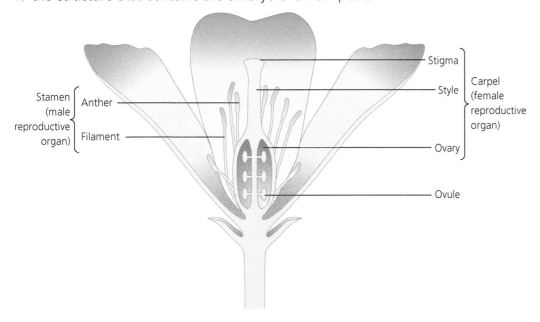

Apply

1 The table below shows the stages that take place during the process of sexual reproduction of a plant. The stages are not in the correct order. Give each statement a number indicating its order in the process.

Reproduction stage	Order
The fertilised egg develops into an embryo.	
The pollen grain forms a pollen tube, which grows down through the style towards the ovary.	
The ovule grows into the seed.	
A pollen grain from a different plant lands on the stigma of a flower.	
The ovary forms the fruit that surrounds the seed.	
The nucleus of the pollen grain containing DNA passes down the tube to fertilise the egg inside the ovule.	

2 Many of the plant structures involved in reproduction are adapted for their function.

a) What is the function of fruit?

b) Give an example of how fruits are adapted for their function.

Extend

1 Give a reason for each of the following features of wind-pollinated (WP) and insect-pollinated (IP) flowers.

a) WP: large quantities of pollen produced. IP: smaller quantities of pollen produced.

b) WP: light, smooth pollen. IP: sticky or spiky pollen.

c) WP: flowers with no scent, nectar or bright petals. IP: flowers often scented, contain nectar and are brightly coloured.

d) WP: feathery or net-like stigma. IP: sticky coating on stigma.

» Asexual reproduction

Worked example

What is the 'gene pool' of a population of organisms and why is it important?

A gene pool is a measure of how much variation there is between organisms of the same species. If the gene pool is wide, it means that there is lots of variation between individual organisms, but if it is narrow there is little variation. The width of the gene pool is important because the wider it is, the better the chances are that the organism can survive a harmful change to its ecosystem (like the introduction of a new predator).

Know

1 Give an example of a plant that uses runners to complete asexual reproduction.

2 During asexual reproduction, plants produce clones. What is a clone?

Apply

1 Describe the differences between asexual and sexual reproduction.

2 The table below shows the stages that take place during the process of asexual reproduction of a plant that uses runners. The stages are not in the correct order. Give each statement a number indicating its order in the process.

Reproduction stage	Order
Small plantlets (or clones of the parent plant) develop on offshoots of the runner.	
The plantlets develop roots that grow into the soil and the clone plant continues to grow.	
A thin, flexible stem-like structure called a runner starts to grow at the base of the plant's stem.	
In this way, the clone plants do not share the same nutrients, sunlight and water that the parent plant needs.	
The runner grows horizontally along the ground.	

3 Describe an advantage of using runners during asexual reproduction instead of plantlets that fall from the parent plant's leaves.

Extend

1 Self-pollination is when pollen from the same plant pollinates a flower.

Cross-pollination occurs when pollen from a different plant pollinates the flowers of a plant. Soybean plants have flowers that can be pollinated by insects when they are open, but can also self-pollinate as the flowers close.

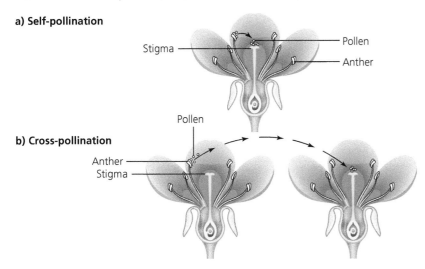

a) Self-pollination

Stigma — Pollen
— Anther

b) Cross-pollination

Pollen
Anther —
Stigma —

a) What type of reproduction will cross-pollination result in?

b) What type of reproduction will self-pollination result in?

c) Explain why having flowers that can do both types of pollination is an advantage.

» Seed dispersal

Worked example

Plants can disperse their seeds using wind, water and animals. What other method can they use to spread their seeds?

Some plants disperse their seeds by ejecting them forcibly from their pods. This shoots the seeds away from the plant.

Know

1 Copy and complete the following paragraph using your knowledge of seed dispersal (use the words in the box).

animals	blown	egestion	float
fruit	lightweight	wind	

Some plants, like dandelions, make seeds that are _____ and can be easily _____ away from the parent by the _____ . Other plants make seeds that can _____ and so can float away from the parent plant on water. Other plants rely on _____ to carry their seeds away. One way they do this is by making _____ that encourages animals to eat them and expel the seeds somewhere else by _____ .

Apply

1 Swan plants, like dandelions, produce seeds that are easily carried large distances by the wind. What is the benefit of producing seeds like this?

2 When we eat a tomato or a strawberry, we eat the seeds as well as the fruit that surrounds them. Why do the seeds not get broken down by our digestive systems?

3 Which method of seed dispersal – wind, water or animal – is most likely to get the seed furthest from the plant? Explain your answer.

Extend

1 One dandelion flower produces approximately 200 seeds. In scientific tests it was shown that about 99.5% of these seeds are carried no more than 1 m from the parent plant, only 0.05% are carried more than 10 m and only 0.014% of seeds are carried more than 1 km!

   a) How many seeds from one flower are transported no more than 1 m?

   b) For every 10 dandelion flowers, one seed is carried more than 10 m. How many dandelion flowers are needed before one seed is transported more than 1 km?

   c) One meadow contains 3000 dandelion flowers. How many seeds from this meadow will be carried more than 1 km from this meadow?

> **Hint**
> One seed is 0.05% of 2000 seeds.

» Artificial reproduction in plants

Extend

1 Wild bananas are nothing like the long, curved, easy-peeling, seedless fruits that many of us enjoy today. Wild bananas are small, oval, have a tough skin and are full of large tough seeds – they would not make a pleasant snack! Our snack bananas have been selectively bred over many years.

   a) Describe how early banana breeders would have chosen and bred wild bananas to produce the fruit we eat today.

   b) Because of the way banana plants have been selectively bred, they are unable to cope with diseases and changes in the weather. Explain why.

2 Selective breeding of important food plants like wheat, rice and maize has helped food production to keep up with global population growth. Describe and explain four desirable characteristics that important food plants have.

3 Why is it important that we keep populations of insect pollinators large and healthy?

4 Selective breeding and the genetic modification of plants can both result in the same outcome: plants with more desirable characteristics. However, many people feel that genetic modification of organisms is wrong or unethical. Use the internet to find out how genetic modification is different to selective breeding.

19 Variation

» Causes of variation

Worked example

Dan and Greg are unrelated, and Dan is taller than Greg. Explain why this variation in their heights could be an example of genetic and environmental variation.

Dan could have inherited genes that make him naturally taller than Greg, so the height difference is an example of genetic variation. But he could also have had a better diet than Greg as he was growing up, so he grew taller; an example of environmental variation.

Know

1 Copy and complete the following paragraph using your knowledge about species and breeding (use the words in the box).

donkey fertile infertile horse species species

A _____ is a group of living things that have more in common with each other than with other groups. Different organisms of one species are able to interbreed and have _____ offspring. Organisms from different species may be able to breed, but their offspring will be _____ . This means that the offspring cannot breed themselves. For example, a mule can never produce offspring because it was made when a _____ and _____ bred, and these animals are not in the same _____ .

Apply

1 For the following examples of variation within a species, state whether the variation is an example of genetic or environmental variation or both:

   a) two giraffes with different heights

   b) one dog with brown eyes and another with pale blue eyes

   c) one person with short hair and another with long hair

   d) one person with naturally red hair and another with naturally blond hair.

2 Imagine that two genetically identical twins grow up separately, and by the time they are adults, one has darker skin than the other. Explain how this could happen.

Extend

1 Like variation, many diseases have genetic or environmental causes, or both. Skin cancer can be caused by living in a very sunny climate (an environmental factor), but some people are genetically more likely to suffer from skin cancer than others. The occurrence of skin cancer in the UK has risen by 360% since the 1970s. Suggest whether genetic or environmental factors are more important in explaining the increase in skin cancer in the UK since 1970. Explain your answer as fully as you can.

» Types of data

Worked example

Give an example of

a) a continuous variation that is caused by genetic and environmental factors
b) a discontinuous variation that is caused by an environmental factor.

For example:
a) height
b) whether a person has tattoos.

Know

1 For each of the following examples of variation, state whether they are continuous or discontinuous variations:

a) weight   **c)** flower colour   **e)** leg length.

b) shoulder width   **d)** blood group

Apply

1 A student decides to survey her classmates' height. She measures each classmate's height and records the following data.

Height of classmate (cm)	Number of classmates with this height
140	1
145	2
148	3
150	4
154	5
157	4
161	3
165	2
168	1

a) What sort of variable is the students' height: continuous or discontinuous?

b) What sort of graph should the student draw to represent her data?

c) Plot a line graph with 'height of classmate' on the *x*-axis and 'number of classmates with this height' on the *y*-axis. Draw a smooth line through your data points.

d) The type of pattern shown here is called a 'bell curve'. Suggest why.

2 In the UK, blood donors donate blood to help other people who need blood. It is important that a patient gets the right blood type. There are eight different blood groups, as shown in the table below.

Blood type	Percentage of donations of that blood type (%)*
O+	36
A+	30
B+	8
AB+	2
O–	13
A–	8
B–	2
AB–	1

*taken from the NHS Give Blood website (www.blood.co.uk/why-give-blood/the-need-for-blood/blood-groups)

a) What type of variable is blood type: continuous or discontinuous? Give a reason for your answer.

b) What sort of graph should we use to represent these data?

c) Draw a graph of these data with 'blood type' on the x-axis, and 'percentage of donations of that blood type' on the y-axis.

Extend

1 In humans, there is a relationship between height and arm span.

a) What sort of data are height and arm span?

b) Predict the relationship between height and arm span.

c) Draw a sketch graph with height on the x-axis and arm span on the y-axis.

» Why is variation important?

Worked example

Selective breeding of pug dogs has resulted in pugs having a very narrow gene pool. Explain how this will affect the breed's ability to survive a change to their environment.

Because there is so little variation in the breed, each pug dog has the same ability to adapt to the change in their environment. So any change that occurs that any one pug dog can't adapt to, will affect all the other dogs in the same way.

> **Hint**
>
> A sketch graph just shows the axes and the line of best fit. It shows the relationship between two factors (or variables) without plotting any data points.

Know

1 Copy and complete the following paragraph using your knowledge of biodiversity and its benefits (use the words in the box).

variation	environment	extinct	gene
inbred	pool	species	

Variation is a measure of the differences within and between _____ .
A similar measure is biodiversity, which is the variety of living things.
We know that species that show a lot of _____ are more able
to cope with change to their _____ , and this is because with
greater variation there is a wider _____ _____ . Species
that can adapt to changes in their environment are less likely to become
_____ than those that can't. Animals, like dogs, that have been bred
to have certain characteristics can often have a very narrow gene pool,
and we say the animal is _____ .

Apply

1 Polar bears are highly adapted to their Arctic environment. They have many characteristics that enable them to live in a very challenging environment.

a) Describe and explain two characteristics of polar bears that help them to live in the Arctic.

b) The Arctic environment is changing rapidly due to global warming and polar bears may become extinct. What will help polar bears to survive this change?

2 Explain how variation in a population of giant tortoises on a remote island would help the population to survive a disease that was brought to the island when a new tortoise was left there.

Hint

If you cannot think of any characteristics, go online and search for 'polar bear adaptations'.

Extend

1 Many animals, like the snow leopard, are critically endangered, meaning they face a very high risk of extinction in the wild. Their populations have decreased significantly.

a) Give an example of how humans have increased the risk of extinction of some animals.

b) As a population of organisms gets smaller, the variation within the population also decreases. What is the significance of this change in variation in terms of survival of the animal?

c) Modern zoos play an important role in the conservation of animals, including breeding programmes. In terms of variation, how can zoos help to conserve animals?

» Evolution

Extend

1 Tawny owls are nocturnal predators. They have grey-brown feathers and the more grey that is present, the better camouflaged the owls are from predators in snowy environments.

   a) What causes variation in feather colour: genetic or environmental factors?

   b) Predict how the colouring of tawny owls might evolve as global warming continues.

   c) Describe how your evolutionary prediction would happen.

2 Great tits are birds that feed on caterpillars, and they have evolved to lay their eggs at the time when caterpillars are most abundant. There is variation in the great tit population in terms of how much they can change the timing of when they lay their eggs. Some can change when they lay their eggs, and others cannot.

   a) As a result of global warming, the time when caterpillars are most abundant is occurring earlier and earlier each year. Which great tits are most likely to survive this change?

   b) Describe how evolution will work in this population of great tits as global warming continues.

   c) The faster global warming occurs, the harder it is for species to adapt and survive. Describe the effect on the population of caterpillars and great tit predators if the great tit population fails to adapt quickly enough to global warming.

20 Human reproduction

» The male and female reproductive systems

Worked example

A sperm cell is an example of a cell with structural adaptations that make it very good at its job. Describe and explain the structural adaptations of a sperm cell.

Sperm cells have a tail that they use to swim toward the egg. They have lots of mitochondria at the base of their tails, which release lots of energy that they use to swim to the egg.

Know

1 Give the name of the structure in the female or male reproductive system

   a) that contains eggs

   b) where a baby develops in a pregnant woman

   c) that carries an egg from the ovary to the uterus, and is where fertilisation occurs

   d) where sperm are produced

   e) that carries sperm out of the male's body.

Apply

1 The following table contains seven statements that describe the process of sexual reproduction and fertilisation in humans. The statements are not in the correct order. Give each statement a number (1–7) showing the correct order.

Statement	Order
During ejaculation, semen enters the woman's vagina near the cervix.	
Many sperm then enter the fallopian tubes (or oviducts).	
Many millions of sperm are released from testicles and mix with a nutrient-rich fluid, forming semen.	
The fertilised egg is wafted into the uterus by ciliated cells that line the fallopian tubes.	
Millions of sperm swim through the cervix and enter the uterus or womb.	
The fertilised egg then settles into the lining of the uterus and develops into a baby.	
One sperm finds and enters the egg and fertilises it.	

Extend

1 The human sperm cell is one of the smallest human cells with the body being only 3 micrometres (μm) wide. The egg cell is the largest human cell, at 120 μm in diameter. 1 μm is 0.001 mm, so 1 mm is 1000 μm.

   a) Normal human hair is approximately 70 μm wide. What is this width of normal human hair in millimetres?

   b) How many times wider is an egg cell compared with a sperm?

   c) We can use an optical microscope to view sperm cells. The size of the image of the sperm will depend on the magnification we use. If the magnification of the eyepiece lens is ×10, and the magnification of the objective lens is ×40, how wide will the body of the sperm appear, in millimetres?

» The menstrual cycle

Worked example

What are hormones, how do they work and why are they important?

Hormones are molecules produced by the body that control lots of important processes like puberty, the menstrual cycle and your blood sugar. They work by being released into the bloodstream from glands in the body. They eventually enter their target organ and cause a change in that organ or how it works.

Know

1 Define the following terms:

   a) ovulation

   b) menstruation

   c) fetus.

Apply

1 Describe what is happening within the female reproductive system at the following times during a typical menstrual cycle of a woman. Use as many key words in your answers as you can.

   a) day 14

   b) days 1–4

   c) days 15–28 (if the egg is not fertilised)

   d) days 15–28 (if the egg is fertilised)

2 Hormones are important biological molecules that control many processes in the human body. Oestrogen and progesterone are two important hormones for the menstrual cycle.

   a) Describe how the levels of oestrogen in a woman's bloodstream affect the menstrual cycle.

   b) Describe how the levels of progesterone in a woman's bloodstream affect the menstrual cycle.

Extend

1 The human egg cell lives for 12–24 hours following ovulation while sperm cells can survive for up to seven days inside a woman's reproductive system.

   a) Suggest the best time during a woman's menstrual cycle to conceive (become pregnant). Explain your answer.

   b) At what time during a woman's menstrual cycle is there the lowest chance of fertilisation taking place?

   c) Why might fertilisation take place if sex takes place on day 10 of the menstrual cycle?

2 During ejaculation, a healthy man will release approximately 1 200 000 sperm cells, while a woman will have usually only released one egg cell. Use your knowledge and understanding, or the internet, to explain why so many sperm cells are released when there is only one egg cell to fertilise.

» Gestation and birth

Worked example

Describe how some mammals, like koalas and kangaroos, have a different gestation to other mammals, like cats, humans and giraffes.

Koalas and kangaroos have short gestation periods. At the end of gestation, the baby koala or kangaroo climbs into its mother's pouch where the rest of its development takes place. Koalas and kangaroos are examples of marsupials.

Know

1 The placenta plays an essential role in the growth and development of a fetus.

   a) What essential substances diffuse through the placenta from the mother to the fetus?

   b) What substances diffuse through the placenta from the fetus to the mother?

Apply

1 Explain why pregnant women are advised not to smoke or drink a lot of alcohol during pregnancy?

2 The following table contains six statements that describe the process of birth in humans. The statements are not in the correct order. Copy the table and put a number beside each statement (1–6) showing the correct order.

Statement	Order
The muscles of the uterus will undergo a series of involuntary contractions.	
The placenta passes from the uterus through the vagina.	
The baby moves from the uterus, through the narrow opening of the cervix, to the vagina.	
The amniotic sac surrounding the baby bursts and releases the amniotic fluid, which passes from the vagina.	
The baby's head is usually born first, followed by the rest of its body.	
The umbilical cord is cut and the baby takes its first breath.	

Extend

1 Smoking during pregnancy can have serious health effects for the fetus. Cigarette smoke contains thousands of harmful chemicals, including nicotine and carbon monoxide.

   a) Carbon monoxide is a toxic gas that can pass through the placenta and that limits how much oxygen red blood cells can carry. Suggest how carbon monoxide may affect the development of a fetus.

   b) Nicotine can limit the diffusion of nutrients across the placenta. How might this affect a developing fetus?

   c) Some people use e-cigarettes instead of cigarettes, as they still contain nicotine (the addictive substance in cigarettes) but fewer toxic chemicals. E-cigarettes are very new. Why can we not be certain that e-cigarettes are safer for pregnant women than normal cigarettes?

» Contraception and fertility

Extend

1 There are several reasons for infertility in women and men. Use your knowledge and understanding to suggest how the following causes of infertility reduce the chances of fertilisation occurring:

   a) damage to a woman's fallopian tubes

   b) irregular ovulation

   c) a thickening of the fluid around the woman's cervix

   d) a low sperm count

   e) low levels of the hormone testosterone in the man.

> **Hint**
> Do you think you could swim in syrup?

> **Hint**
> You may need to use the internet to find out how testosterone affects the male reproductive system.

2 There are many different forms of contraception including condoms, the contraceptive pill and the diaphragm. The following table gives information about two other forms of contraception for women: the contraceptive patch, which is worn on the arm, and the IUD (intrauterine device), which is implanted inside the woman's uterus.

Use the information in the table to decide which is the best form of contraception. Explain your reasoning.

Contraceptive patch	IUD
Greater than 99% effective	Greater than 99% effective
Lasts for one week before needing to be replaced.	Lasts from five to ten years before needing to be replaced.
Easy to apply.	Requires a trained nurse or GP to implant the device.
May take up to one week to start working.	Works immediately.
Can cause sickness and headaches.	Can be painful to fit.
May reduce the risk of ovarian cancer.	In fewer than one in 1000 women, it can damage the uterus.
Does not protect against sexually transmitted infections (STIs).	Does not protect against STIs.
Not suitable for all women.	Suitable for most women.
Can make the menstrual cycle more regular.	Does not affect the menstrual cycle.

3 Before beginning fertility treatments like IVF, both the man and woman undergo a series of fertility tests.

For each of the following fertility tests, explain what the test investigates or checks, and how this factor may affect the fertility of the man or woman.

a) A sperm test for the man

b) A blood test for the woman

c) An ultrasound scan of the woman's reproductive organs

> **Hint**
>
> Ultrasound is a medical imaging technique used to see internal organs and fetuses.

Index